Lecture Notes in Computer Science 2570

Edited by G. Goos, J. Hartmanis, and J. van Leeuwen

T0224692

Michael Jünger
Gerhard Reinelt
Giovanni Rinaldi (Eds.)

Combinatorial Optimization – Eureka, You Shrink!

Papers Dedicated to Jack Edmonds

5th International Workshop
Aussois, France, March 5-9, 2001
Revised Papers

 Springer

Series Editors

Gerhard Goos, Karlsruhe University, Germany
Juris Hartmanis, Cornell University, NY, USA
Jan van Leeuwen, Utrecht University, The Netherlands

Volume Editors

Michael Jünger
Institut für Informatik
Universität zu Köln
Pohligstr. 1, 50969 Köln, Germany
E-mail: mjuenger@informatik.uni-koeln.de

Gerhard Reinelt
Institut für Informatik
Universität Heidelberg
Im Neuenheimer Feld 368, 69120 Heidelberg, Germany
E-mail: gerhard.reinelt@informatik.uni-heidelberg.de

Giovanni Rinaldi
Istituto di Analisi dei Sistemi ed Informatica "Antonio Ruberti"
CNR
viale Manzoni 30, 00185 Rome, Italy
E-mail: rinaldi@iasi.rm.cnr.it

Cataloging-in-Publication Data applied for

A catalog record for this book is available from the Library of Congress.

Bibliographic information published by Die Deutsche Bibliothek
Die Deutsche Bibliothek lists this publication in the Deutsche Nationalbibliografie;
detailed bibliographic data is available in the Internet at <http://dnb.ddb.de>.

CR Subject Classification (1998): G.1.6, G.2.1, F.2.2, I.3.5

ISSN 0302-9743
ISBN 3-540-00580-3 Springer-Verlag Berlin Heidelberg New York

Springer-Verlag Berlin Heidelberg New York
a member of BertelsmannSpringer Science+Business Media GmbH

http://www.springer.de

© Springer-Verlag Berlin Heidelberg 2003
Printed in Germany

Typesetting: Camera-ready by author, data conversion by PTP-Berlin, Stefan Sossna e. K.
Printed on acid-free paper   SPIN: 10872263   06/3142   5 4 3 2 1 0

Preface

A legend says that Jack Edmonds shouted "Eureka – you shrink!" when he found a good characterization for matching (and the matching algorithm) in 1963, the day before his talk at a summer workshop at RAND Corporation with celebrities like George Dantzig, Ralph Gomory, and Alan Hoffman in the audience. During Aussois 2001, Jack confirmed: "'Eureka – you shrink!' is really true, except that instead of 'Eureka' it maybe was some less dignified word."

Aussois 2001 was the fifth in an annual series of workshops on combinatorial optimization that are alternately organized by Thomas Liebling, Denis Naddef, and Laurence Wolsey – the initiators of this series – in even years and the editors of this book in odd years (except 1997).

We decided to dedicate Aussois 2001 to Jack Edmonds in appreciation of his groundbreaking work that laid the foundations of much of what combinatorial optimizers have done in the last 35 years. Luckily, Jack is a regular participant of the Aussois workshops and, as ever, he cares a lot for young combinatorial optimizers who traditionally play a major rôle in the Aussois series.

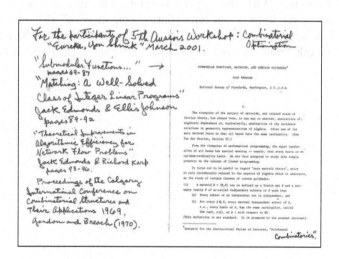

Fig. 1. Handout for the Aussois 2001 participants

Highlights of Aussois 2001 included a special session entitled "Eureka – you shrink!" in honor of Jack and a special lecture by Jack on "Submodular Functions, Matroids, and Certain Polyhedra" that closed the workshop.

In this book, we give an account of the "Eureka – you shrink!" session as well as reprints of three hardly accessible papers that Jack gave as a handout to the Aussois 2001 participants and that were originally published in the *Proceedings of the Calgary International Conference on Combinatorial Structures and Their Applications 1969*, Gordon and Breach (1970) – newly typeset by the editors and reprinted by permission.

We are happy that 13 speakers of Aussois 2001 agreed to dedicate revisions of the papers they presented during the workshop to Jack Edmonds. Their contributions have made this book possible.

As organizers and editors we would like to thank Jack Edmonds, the authors, Miguel Anjos for transcribing Bill Pulleyblank's speech, and, in particular, Bill Pulleyblank for support that went well beyond the contribution you will find in the "Eureka – you shrink!" chapter!

When Jack gave an interview for the Kitchener-Waterloo Record of May 30, 1985, on the occasion of the award of the John von Neumann Theory Prize to him a month earlier, he said: "I hit it lucky by putting a poetic title on a paper that was mathematically a hit," referring to his seminal paper entitled "Paths, Trees, and Flowers." With the help of Matthias Elf, we found an artist whose proposal for a cover illustration immediately convinced all three of us: People who know Jack's work will have the right association without reading the words. A special "Thank You!" to Thorsten Felden who contributed his own hommage à Jack!

January 2003

Cologne   Michael Jünger
Heidelberg   Gerhard Reinelt
Rome   Giovanni Rinaldi

Participants

Aardal, Karen (University of Utrecht)
Ahr, Dino (University of Heidelberg)
Amaldi, Edoardo (Politecnico di Milano)
Békési, Jozsef (University of Szeged)
Bixby, Robert E. (Rice University, Houston)
Buchheim, Christoph (University of Cologne)
Cameron, Kathie (Wilfried Laurier University, Waterloo)
Caprara, Alberto (University of Bologna)
Edmonds, Jack (University of Waterloo)
Elf, Matthias (University of Cologne)
Euler, Reinhardt (University of Brest)
Evans, Lisa (Georgia Institute of Technology, Atlanta)
Farias, Ismael de (University at Buffalo)
Feremans, Corinne (University of Brussels)
Fischetti, Matteo (University of Padova)
Fleischer, Lisa (Columbia University, New York)
Fortz, Bernard (University of Brussels)
Galambos, Gábor (University of Szeged)
Gruber, Gerald (Carinthia Tech Institute, Villach)
Hemmecke, Raymond University of Duisburg)
Johnson, Ellis (Georgia Institute of Technology, Atlanta)
Jünger, Michael (University of Cologne)
Kaibel, Volker (University of Technology, Berlin)
Labbé, Martine (University of Brussels)
Lemaréchal, Claude (INRIA Rhône-Alpes)
Letchford, Adam (Lancaster University)
Liers, Frauke (University of Cologne)
Lodi, Andrea (University of Bologna)
Lübbecke, Marco (University of Braunschweig)
Luzzi, Ivan (University of Padova)
Maffioli, Francesco (Politecnico di Milano)
Martin, Alexander (University of Technology, Darmstadt)
Maurras, Jean François (Université de la Mediterrané, Marseille)
Meurdesoif, Philippe (INRIA Rhône-Alpes)

Möhring, Rolf H. (University of Technology, Berlin)

Monaci, Michele (University of Bologna)

Mutzel, Petra (University of Technology, Vienna)

Naddef, Denis (ENSIMAG, Montbonnot Saint Martin)

Nemhauser, George (Georgia Institute of Technology, Atlanta)

Nguyen, Viet Hung (Université de la Mediterrané, Marseille)

Ortega, Francois (CORE, Louvain-la-Neuve)

Oswald, Marcus (University of Heidelberg)

Percan, Merijam (University of Cologne)

Perregard, Michael (Carnegie Mellon University, Pittsburgh)

Pulleyblank, William (IBM Yorktown Heights)

Reinelt, Gerhard (University of Heidelberg)

Remshagen, Anja (University of Texas at Dallas)

Rendl, Franz (University of Klagenfurt)

Richard, Jean Philippe (Georgia Institute of Technology, Atlanta)

Riis, Morton (University of Aarhus)

Rinaldi, Giovanni (IASI-CNR Rome)

Rote, Günter (Free University of Berlin)

Salazar-González, Juan-José (University of La Laguna, Tenerife)

Schultz, Rüdiger (University of Duisburg)

Skutella, Martin (University of Technology, Berlin)

Spille, Bianca (EPFL-DMA Lausanne)

Stork, Frederik (University of Technology, Berlin)

Toth, Paolo (University of Bologna)

Uetz, Marc (University of Technology, Berlin)

Vandenbusse, Dieter (Georgia Institute of Technology, Atlanta)

Verweij, Bram (Georgia Institute of Technology, Atlanta)

Weismantel, Robert (University of Magdeburg)

Wenger, Klaus (University of Heidelberg)

Woeginger, Gerhard (University of Twente)

Wolsey, Laurence (CORE, Louvain-la-Neuve)

Zimmermann, Uwe (University of Technology, Braunschweig)

Table of Contents

"Eureka – You Shrink!"

Surprise Session for Jack Edmonds

In the evening of March 7, 2001, there was a surprise session for Jack Edmonds entitled "Eureka – You Shrink!".

Fig. 1. Surprise Session for Jack

Bill Pulleyblank acted as chairman, master of ceremonies, and main speaker. Here is a transcript of his speech.

William R. Pulleyblank:

Welcome to this session. We planned this to be the pinnacle of this exciting week in Aussois, because, as the organizers said, we are holding this meeting in honour of Jack's 66th birthday, which will happen April 5th of this year. What I wanted to do though was to talk a bit about Jack's early career, because I think there are many of us now who don't realize some of the really big ideas that he came up with at that time and the impact that they subsequently had.

Now, let me try to put things in perspective. I'm going to start around 1960. I was in junior high school, so I don't actually remember these things happening. But, I have this on good authority.

M. Jünger et al. (Eds.): Combinatorial Optimization (Edmonds Festschrift), LNCS 2570, pp. 1–10, 2003.
© Springer-Verlag Berlin Heidelberg 2003

How many of you were born after 1960? Put up your hands with me. (laughter) OK, Keep them up. (more laughter) You guys too were born after that? (yet more laughter) Like they say in the newspaper business, hey Jack, that stuff you did had legs, its keeps on going!

Fig. 2. William R. Pulleyblank

In 1960, computers were in a pretty early state. Blue Gene had not even been conceived of, let alone started to be built. Do you remember? THE algorithm people knew best was the simplex algorithm that solved linear programming. It had been created in 1947, and a real code was created by Orchard Hays in the early fifties.

Now, late in the 50s, I don't know if you remember but there was an exciting development. Ralph Gomory published an algorithm that solved integer programming problems. Now, remember, the simplex algorithm for linear programming, yes, in theory it could take a lot of iterations, but it never did. It worked. It was a great algorithm and then Gomory had this brilliant idea of making what looked like a superficial modification of the simplex algorithm and it now could solve integer programming problems. By the way, he proved that it was finite, just as it had been proved for the simplex algorithm.

Problems were finite or infinite, and once a problem was known to be finite there were no algorithmic questions to be asked because it was all over. I remember when I took my first combinatorics class from the distinguished combinatorialist Eric Milner, and there was a point where we were talking about a theorem, and I said "how would you find one of these?". And he looked at me with a kind look, but the sort of a look a parent gives a child when he says something sort of stupid. He simply said "But Bill, it's finite", and I said "Oh! of course, it's finite".

Fig. 3. "It's finite!"

Now, some of you may not know, but Jack's first paper had nothing to do with algorithms, it had to do with embeddings of graphs and orientations of surfaces. This is a paper which was quite well known subsequently in the community of people who worked in this domain. In fact, I've met people in Cologne who actually knew Jack because of that paper. I asked what they thought of his later work and they said "I haven't seen anything in years" because they were sticking to the embedding areas. And they were always sorry, Jack, that you got out of it.

Things got really interesting when we got into the 60s. In this world where all everybody cared about was finiteness. Jack proceeded to do several things. First of all, he defined this notion of a good algorithm – one whose running time only grew polynomially with the input size. Now, that it itself was a significant intellectual step, but what he proceeded to then was knock off a bunch of problems and show how they could be solved in polynomial time to give substance to this idea.

You see before you what I shall call exhibit A. This is an authentic reprint of "Paths, Trees and Flowers" from the Canadian Journal of Mathematics, volume 17, 1965. You can all see the autograph on the front. This is worth more than a DiMaggio autograph because I understand there are far fewer Edmonds' autographs than there are DiMaggio's. (pause) Joe DiMaggio was a baseball player. (laughter)

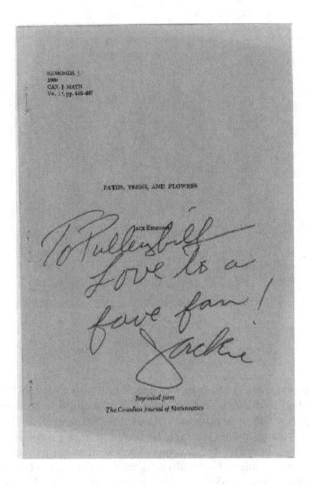

Fig. 4. Exhibit A

This is the paper in which Jack described how to find maximum matchings in graphs in $O(n^4)$, because for Jack it wasn't so important if it was n^3 or n^4, and in fact when he discussed this he said "Oh! yes, I guess you can do it in n^3", but that wasn't really the point. The point was, it was better than finite, it was better than exponential.

Now, if you haven't read it already, at some point I might loan it to you. But what I would like to do now, if you would just let me have a moment, is read from Section 2 of this paper. This is one of the most interesting sections because this is what laid out the agenda for forty years of research (so far).

Section 2 is entitled "Digression". Was that the first time you ever digressed, Jack? (laughter)

> *An explanation is due on the use of the words 'efficient algorithm'. First, what I present is a conceptual description of an algorithm, and not a particular formalized algorithm or code.*

First off, Jack wanted to make it clear that he didn't have to write FORTRAN or anything like that, but he had this abstract idea of an algorithm.

> *For practical purposes, computational details are vital. However, my purpose is only to show as attractively as I can that there is an efficient algorithm. According to the dictionary, efficient means 'adequate in operational performance'. This is roughly the meaning I want, in the sense that it is conceivable for maximum matching to have no efficient algorithm. Perhaps a better word is 'good'. I am claiming as a mathematical result the existence of a good algorithm for finding a maximum cardinality matching in a graph. There is an obvious finite algorithm, but that algorithm increases in difficulty exponentially with the size of the graph. It is by no means obvious whether or not there exists an algorithm whose difficulty increases only algebraically with the size of the graph.*

Pretty explicit, isn't it? And that was laying out this agenda where polynomiality, and the importance of it, and the fact that this was a mathematically substantive question, was an issue which could be raised and a problem which could be posed.

Not surprisingly, when the paper appeared, the reaction of the world was pretty much what you would expect. People kept on doing what they were doing before.

Jack, I never asked you if you were discouraged, but you sure kept the papers coming out. In that period, there were a number of papers and for a while, in fact, it was funny, because if you read "Paths, Trees and Flowers", which is (maximum) cardinality matching in a nonbipartite graph, you finally get towards the end, and when Jack is throwing in the goodies at the end, he says "Oh! by the way, if you want to solve the weighted problem, here's what you would have to do. You write this set of inequalities, put it in there. But I'm not going to talk about it here, that's in the next paper" which came out in "Maximum matching and a polyhedron with 0,1-vertices". which appeared in the Bulletin of the National Bureau of Standards, around the same time, and which actually knocked off the weighted version of that problem. And when he wrapped up the last section of that paper, he said "Oh!, and by the way, if it's capacities and b-matchings you're interested in, here's how all the results generalize for this", and he just tossed it off. So it went on, this sequence of solving more and more general matching problems.

Now roughly at the same time, this whole area of matroids suddenly leapt into prominence for optimization. Matroids had existed for a long time. Does anybody know what year they appeared in the literature for the first time? In 1935 Hassler Whitney wrote a paper about matroids which was capturing the abstract properties of linear independence. Beautiful paper. And various people had gone on studying the properties of these algebraic structures. That's the year Jack was born, right? 1935? Is that true? In 1935? 1932? Two years? Jack, you were born in 1935! (laughter) Yeah, that wouldn't make sense, would it? I think it was 1935, but here's what was interesting. (Jack: At that time, I worked on the subject, it was independently.) Good! Let's say in the early 30s.

Now, you see the thing about matroids was that they were this abstract structure, and I guess there were actually several people who somehow had this idea of optimizing over matroids with the greedy algorithm. And, Jack, of course, knew how to do that. But he was now beginning to strut his stuff with this new paradigm he had developed, which was the idea of polyhedral combinatorics. You see, in fact when Jack first described the matching polytope, he would write the degree constraints, then he would put the blossom inequalities in. There were a lot of people at that point who said "Ah! That's really a bad idea", and I remember people saying "You know Jack, they don't make computers big enough to store all those inequalities." And Jack said "Yes, but you don't need to store them explicitly. You can generate them, and use them when you need them". They said "I don't get it", so there was a bit of a problem on that point. But it was this idea that you could add this large set of inequalities to a small set of inequalities that define an integer program, and from there go on to the algorithm and to proving optimality, and the whole NP characterization idea which was in there.

Now I guess the thing that to me was particularly remarkable about this was the stunning sequence of papers that followed including matroid partition, matroid intersection – much more complex combinatorial structures which came from this very simple notion of a matroid. Then, damned if Jack didn't go ahead and solve the corresponding optimization problems.

Now there was one problem that we always ran into with the whole matroid paradigm. People would say "Now, how do you know when a set is independent in a matroid?" and again get into these questions of defining oracles. At one point Jack said, again I wasn't there at the time, but I can just imagine how Jack finally said it, "Well, I'll give them something simple they can understand", so in Mathematics of the Decision Sciences, there was a paper on what Jack called "branchings". Optimum branchings was to be an accessible example of matroid intersection, so people would actually see this stuff working. And I do remember Jack telling me at a point later that he was a little disappointed that it took so many pages to write it down, even though it seemed so simple. But the branching thing came out as an elegant concrete example of matroid intersections.

Another point that I would like to stress here, this whole idea of complexity of integer calculation was discovered by Jack. There's a paper called "Systems of distinct representatives and linear algebra" which contains the famous line: "Gaussian elimination is not a good algorithm". And, I can remember when I first read that line, I thought "I don't get it. You clearly only have to perform a polynomial number of steps to reduce a matrix, so what's the deal?" And of course what Jack had observed was that the numbers can get big. And the fact that you have to pay attention to the size of the numbers was something that again became extremely influential as people began studying these areas and going on from there.

Partition matroids was another class of matroids Jack invented. It was beginning to get a buzz around it. In 1969, the University of Calgary hosted a massive combinatorial meeting, two weeks long, and basically everybody who was anybody in combinatorics at that time was invited. I had just finished my Master's degree the week before that meeting was to take place, and my Master's thesis actually was on matroids. What my supervisor Eric Milner had done is given me stacks of papers to read on matroids. Bob, you and I are probably the only people who tried to read the Tutte paper and I think you made it further than I did. But there was all this excitement about matroids and all these matroid optimization results were coming from Jack. That was one of the reasons why they invited Jack to come and give a series of lectures at this big Calgary meeting on his research. And, in some sense, you can say this was a coming-out party. It was an endorsement of Jack's views, it gave him a platform to present this material. It was an amazing series of lectures because I can remember that, at this point, Jack had developed the whole idea of submodular functions and polymatroids. This was matroids generalized to a much broader concept, and every so often he would throw in one of those combinatorial realizations. As someone who sat and took notes on those lectures, I remember I was exhausted at the end just trying to keep up with it.

So when I think of the sort of work that Jack did in that stage, not only did he define and promote this whole concept of good algorithms and good characterizations, he solved the matching problem, he did the same for matroid intersections, he did the same for submodular functions, he laid the basis for what became arithmetic complexity.

I think at one point I said to Jack: "How come you got there when all the good stuff was there?". But I think it's probably fairer to say that all this good stuff was there because Jack made it there. He had this vision of what could be done, he could see how it could go, and for those of you who just met him, you may not be aware of it, maybe I am the only one who has noticed this: Jack has a slightly stubborn streak.

I think that served Jack incredibly well, as he created this whole agenda describing where the world of combinatorial optimization should go. I think it's been remarkably successful.

Now, what we want to do at this point is to have a few comments by some of the people who knew Jack somewhat earlier in his career. I think, Jack, the technical term for this is a "roast".

At this point, Ellis Johnson, George Nemhauser, Bob Bixby, Jean-François Maurras, Denis Naddef, and Kathie Cameron successively joined Bill Pulleyblank and all presented reminiscences and a number of entertaining stories, accompanied by lot of laughter and interaction with the audience.

Finally, Jack Edmonds took the stage. We cite from his speech:

> *I should say that I didn't expect to give this talk tonight. These guys,*
> *I saw a session for Edmonds on the board, and then they said to me,*

Fig. 5. "I remember ..."

Fig. 6. Jack Edmonds

"Hey, would you give a few words about the early days". So here I am now. (laughter) It occurred to me that "Yeah, sure, their idea of a special session for Jack is to start him talking." (laughter)

And talk he did, sharing his personal recollections of his early career with the audience. We recommend "A Glimpse of Heaven" [6] which contains Jack's reminiscences on this period of his life, for those readers who were not at Aussois to enjoy it live.

The session closed with a standing ovation for Jack and a presentation to Jack: a poster brought to Aussois by Jean-François Maurras showing Jack in the 70s and signed by the participants of Aussois 2001.

Fig. 7. Jack Edmonds in the 70s

References

1. J. Edmonds (1960) A combinatorial representation for polyhedral surfaces, *Notices American Mathematical Society* 7 (1960), 643
2. J. Edmonds (1965) Paths, trees and flowers, *Can. J. Math.* 17, 449–467
3. J. Edmonds (1965) Maximum matching and a polyhedron with 0,1-vertices, *J. Res. Nat. Bur. Standards Sect. B* 69B, 125–130
4. J. Edmonds (1967) Systems of distinct representatives and linear algebra, *J. Res. Nat. Bur. Standards Sect. B* 71B, 241–245
5. J. Edmonds (1968) Matroid partition, *Mathematics of the Decision Sciences, Part I*, (Proceedings of the Fifth Summer Seminar on the Mathematics of the Decision Sciences, Stanford University, Stanford, CA, 1967), 335–345
6. J. Edmonds (1991) A Glimpse of Heaven in: History of Mathematical Programming: A Collection of Personal Reminiscences (J.K. Lenstra, A.H.G. Rinnoy Kan & A. Schrijver (eds.), North-Holland, 32–54

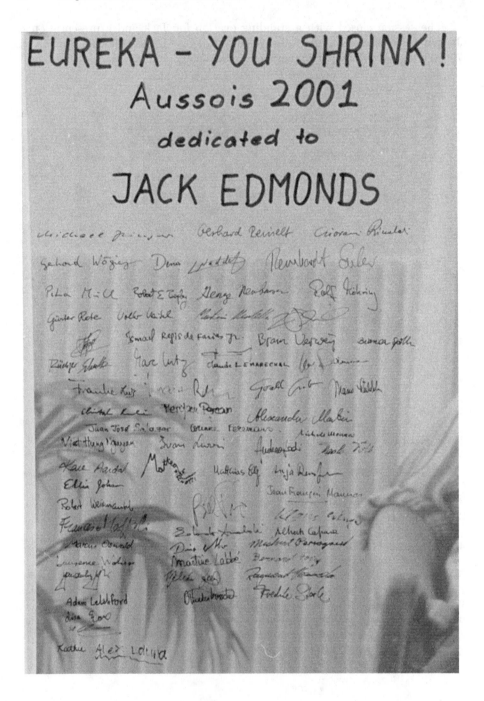

Fig. 8. Signatures of Participants of Aussois 2001

Submodular Functions, Matroids, and Certain Polyhedra*

Jack Edmonds

National Bureau of Standards, Washington, D.C., U.S.A.

I

The viewpoint of the subject of matroids, and related areas of lattice theory, has always been, in one way or another, abstraction of algebraic dependence or, equivalently, abstraction of the incidence relations in geometric representations of algebra. Often one of the main derived facts is that all bases have the same cardinality. (See Van der Waerden, Section 33.)

From the viewpoint of mathematical programming, the equal cardinality of all bases has special meaning — namely, that every basis is an optimum-cardinality basis. We are thus prompted to study this simple property in the context of linear programming.

It turns out to be useful to regard "pure matroid theory", which is only incidentally related to the aspects of algebra which it abstracts, as the study of certain classes of convex polyhedra.

(1)   A *matroid* $M = (E, F)$ can be defined as a finite set E and a nonempty family F of so-called *independent* subsets of E such that

  (a) Every subset of an independent set is independent, and

  (b) For every $A \subseteq E$, every maximal independent subset of A, i.e., every *basis* of A, has the same cardinality, called the *rank*, $r(A)$, of A (with respect to M).

(This definition is not standard. It is prompted by the present interest).

(2)   Let $\mathbb{R}_E$ denote the space of real-valued vectors $x = [x_j]$, $j \in E$. Let $\mathbb{R}_E^+ = \{x : 0 \le x \in \mathbb{R}_E\}$.

(3)   A *polymatroid* P in the space $\mathbb{R}_E$ is a compact non-empty subset of $\mathbb{R}_E^+$ such that

  (a) $0 \le x^0 \le x^1 \in P \implies x^0 \in P$.

  (b) For every $a \in \mathbb{R}_E^+$, every maximal $x \in P$ such that $x \le a$, i.e., every *basis* x of a, has the same sum $\sum_{j \in E} x_j$, called the rank, $r(a)$, of a (with respect to P).

---

* Synopsis for the Instructional Series of Lectures, "Polyhedral Combinatorics".

M. Jünger et al. (Eds.): Combinatorial Optimization (Edmonds Festschrift), LNCS 2570, pp. 11–26, 2003.
© Springer-Verlag Berlin Heidelberg 2003

Here *maximal x* means that there is no $x' > x$ having the properties of x.

(4)   A polymatroid is called *integral* if (b) holds also when a and x are restricted to being integer-valued, i.e., for every integer-valued vector $a \in \mathbb{R}_E^+$, every maximal integer-valued x, such that $x \in P$ and $x \le a$, has the same sum $\sum_{j \in E} x_j = r(a)$.

(Sometimes it may be convenient to regard an *integral polymatroid* as consisting only of its integer-valued members).

(5)   Clearly, the 0–1 valued vectors in an integral polymatroid are the "incidence vectors" of the sets $J \in F$ of a matroid $M = (E, F)$.

II

(6)   Let f be a real-valued function on a lattice L. Call it a β_0-function if

(a) $f(a) \ge 0$ for every $a \in K = L - \{\emptyset\}$;
(b) is non-decreasing: $a \le b \implies f(a) \le f(b)$; and
(c) submodular:

$$f(a \vee b) + f(a \wedge b) \le f(a) + f(b)$$

for every $a \in L$ and $b \in L$.
(d) Call it a β-function if, also, $f(\emptyset) = 0$. In this case, f is also subadditive, i.e., $f(a \vee b) \le f(a) + f(b)$.

(We take the liberty of using the prefixes *sub* and *super* rather than "upper semi" and "lower semi". *Semi* refers to either. The term *semi-modular* is taken from lattice theory where it refers to a type of lattice on which there exists a semimodular function f such that if a is a maximal element less than element b then $f(a) + 1 = f(b)$. See [1].)

(7)   For any $x = [x_j] \in \mathbb{R}_E$, and any $A \subseteq E$, let $x(A)$ denote $\sum_{j \in A} x_j$.

(8)   **Theorem.**   *Let L be a family of subsets of E, containing E and $\emptyset$, and closed under intersections, $A \cap B = A \wedge B$. Let f be a β_0-function on L. Then the following polyhedron is a polymatroid:*

$$P(E, f) = \{x \in \mathbb{R}_E^+ : x(A) \le f(A) \text{ for every } A \in L - \emptyset = K\}.$$

Its rank function r is, for any $a = [a_j] \in \mathbb{R}_E^+$,

$$r(a) = \min \left(\sum_{j \in E} a_j z_j + \sum_{A \in K} f(A) y_A \right)$$

where the z_j's and y_A's are 0's and 1's such that for every $j \in E$,

$$z_j + \sum_{j \in A \in K} y_A \geq 1.$$

Where $f(\emptyset) \geq 0$, only one non-zero y_A is needed.

Where f is integer-valued, $P(E, f)$ is an integral polymatroid.

(9) **Theorem.** *A function f of all sets $A \subseteq E$ is itself the rank function of a matroid $M = (E, F)$ iff it is an integral β-function such that $f(\{j\}) = 1$ or 0 for every $j \in E$. Such an f determines M by:*

$$J \in F \iff J \subseteq E \text{ and } |J| = f(J).$$

(10) For any $a = [a_j] \in \mathbb{R}_E^+$ and $b = [b_j] \in \mathbb{R}_E^+$, let $a \vee b = [u_j] \in \mathbb{R}_E^+$ and $a \wedge b = [v_j] \in \mathbb{R}_E^+$, where

$$u_j = \max(a_j, b_j) \quad \text{and} \quad v_j = \min(a_j, b_j).$$

(11) **Theorem.** *The rank function $r(a)$, $a \in \mathbb{R}_E^+$, for any polymatroid $P \subset \mathbb{R}_E^+$, is a β-function on $\mathbb{R}_E^+$ relative to the above $\vee$ and $\wedge$.*

(12) For any $x = [x_j] \in \mathbb{R}_E^+$ and any $A \subseteq E$, let $x/A = [(x/A)_j] \in \mathbb{R}_E^+$ denote the vector such that $(x/A)_j = x_j$ for $j \in A$, and $(x/A)_j = 0$ for $j \notin A$.

(13) Given a polymatroid $P \subset \mathbb{R}_E^+$, let $\alpha \in \mathbb{R}_E^+$ be an integer-valued vector such that $x < \alpha$ for every $x \in P$. Where r is the rank function of P, let $f_P(A) = r(\alpha/A)$ for every $A \subseteq E$.

Let $L_E = \{A : A \subseteq E\}$. Clearly, by (11), f_P is a β-function on L_E. Furthermore, if P is integral, then f is integral.

(14) **Theorem.** *For any polymatroid $P \subset \mathbb{R}_E^+$,*

$$P = P(E, f_P).$$

Thus, all polymatroids $P \in \mathbb{R}_E^+$ are polyhedra, and they correspond to certain β-functions on L_E.

Theorem 8 provides a useful way of constructing matroids which is quite different from the usual algebraic constructions.

(15) For any given integral β_0-function f as in (8), let a set $J \subseteq E$ be a member of F iff for every $A \in K = L - \{\emptyset\}$, $|J \cap A| \leq f(A)$. In particular, where $L = L_E$, let a set $J \subseteq E$ be a member of F when for every $\emptyset \neq A \subseteq J$, $|A| \leq f(A)$. Then (8) implies that $M = (E, F)$ is a matroid, and gives a formula for its rank function in terms of f. (This generalizes a construction given by Dilworth [1]).

III

In this section, K will denote $L_E - \{\emptyset\} = \{A : \emptyset \neq A \subseteq E)$.

(16)   Given any $c = \{c_j\} \in \mathbb{R}_E$, and given a β-function f on L_E, we show how to solve the linear program:

$$\text{maximize } c \cdot x = \sum_{j \in E} c_j x_j \text{ over } x \in P(E, f).$$

(17)   Let $j(1)$, $j(2)$, ... be an ordering of E such that

$$c_{j(1)} \geq c_{j(2)} \geq \cdots c_{j(k)} > 0 \geq c_{j(k+1)} \geq \cdots$$

(18)   For each integer i, $1 \leq i \leq k$, let

$$A_i = \{j(1), j(2), \dots, j(i)\}.$$

(19)   **Theorem.** (The Greedy Algorithm).   *$c \cdot x$ is maximized over $x \in P(E, f)$ by the following vector x^0:*

$$x^0_{j(1)} = f(A_1);$$
$$x^0_{j(i)} = f(A_i) - f(A_{i-1}) \quad \text{for } 2 \leq i \leq k;$$
$$x^0_{j(i)} = 0 \qquad\qquad\quad \text{for } k < i \leq |E|.$$

(There is a well-known non-polyhedral version of this for graphs, given by Kruskal [9]. A related theorem for matroids is given by Rado [15]).

The dual *l.p.* is to minimize

$$f \cdot y = \sum_{A \in K} f(A) y(A) \text{ where}$$

(20) $y(A) \geq 0$; and for every $j \in E$, $\sum_{j \in A} y(A) \geq c_j$.

(21)   **Theorem.**   *An optimum solution, $y^0 = [y^0(A)]$, $A \in K$, to the dual l.p. is*

$$y^0(A_i) = c_{j(i)} - c_{j(i+1)} \qquad\qquad \text{for } 1 \leq i \leq k - 1;$$
$$y^0(A_k) = c_{j(k)}; \quad \text{and} \quad y^0(A) = 0 \quad \text{for all other } A \in K.$$

(22)   **Theorem.** Corollary to (19).   *The vertices of the polyhedron $P(E, f)$ are precisely the vectors of the form x^0 in (19) for some sequence $j(1), j(2), \dots, j(k)$.*

(23)   Where f is the rank function of a matroid $M = (E, F)$, (9) and (22) imply that the vertices of $P(E, f)$ are precisely the incidence vectors of the members of F, i.e., the independent sets of M. Such a $P(E, f)$ is called a *matroid polyhedron*.

(24)   Let f be a β-function on L_E. A set $A \in L_E$ is called f-*closed* or an f-*flat*, when, for any $C \in L_E$ which properly contains A, $f(A) < f(C)$.

(25)   **Theorem.**   *If A and B are f-closed then $A \cap B$ is f-closed.*

(In particular, for the f of (9), the f-flats form a "geometric" or "matroid" lattice.)

Proof: Suppose that C properly contains $A \cap B$. Then either $C \not\subseteq A$ or $C \not\subseteq B$. Since f is non-decreasing we have $f(A \cap B) \leq f(A \cap C)$ and $f(A \cap B) \leq f(B \cap C)$. Thus, since f is submodular, we have either

$$0 < f(A \cup C) - f(A) \leq f(C) - f(A \cap C) \leq f(C) - f(A \cap B), \quad \text{or}$$
$$0 < f(B \cup C) - f(B) \leq f(C) - f(B \cap C) \leq f(C) - f(A \cap B).$$

(26)   A set $A \in K$ is called f-*separable* when

$$f(A) = f(A_1) + f(A_2)$$

for some partition of A into non-empty subsets A_1 and A_2. Otherwise A is called f-inseparable.

(27)   **Theorem.**   *Any $A \in K$ partitions in only one way into a family of f-inseparable sets A_i such that $f(A) = \sum f(A_i)$. The A_i's are called the f-blocks of A.*

If a polyhedron $P \subset \mathbb{R}_E$ has dimension equal to $|E|$ then there is a unique minimal system of linear inequalities having P as its set of solutions. These inequalities are called the *faces* of P.

It is obvious that a polymatroid $P \subset \mathbb{R}_E^+$ has dimension $|E|$ if and only if, where f is the β-function which determines it, and set $\emptyset$ is f-closed. It is obvious that inequality $x(A) \leq f(A)$, $A \in K$, is a face of $P(E, f)$ only if A is f-closed and f-inseparable.

(28)   **Theorem.**   *Where f is a β-function on L_E such that the empty set is f-closed, the faces of polymatroid $P(E, f)$ are: $x_j \geq 0$ for every $j \in E$; and $x(A) \leq f(A)$ for every $A \in K$ which is f-closed and f-inseparable.*

IV

(29)   Let each V_p, $p = 1$ and 2, be a family of disjoint subsets of H. Where $[a_{ij}]$, $i \in E$, $j \in E$, is the 0–1 incidence matrix of $V_1 \cup V_2 = H$, the following *l.p.* is known as the Hitchcock problem.

(30)   Maximize $c \cdot x = \sum_{j \in E} c_j x_j$, where

(31) $x_j \geq 0$ for every $j \in E$, and $\sum_{j \in E} a_{ij} x_j \leq b_i$ for every $i \in H$.

The dual $l.p.$ is

(32)   Minimize $b \cdot y = \sum b_i y_i$, where

(33) $y_i \geq 0$ for every $i \in H$, and $\sum_{i \in H} a_{ij} y_i \geq c_j$ for every $j \in E$.

Denote the polyhedron of solutions of a system Q by $P[Q]$.

The following properties of the Hitchcock problem are important in its combinatorial use.

(34)   **Theorem.**   (a) *Where the b_i's are integers, the vertices of $P[(31)]$ are integer-valued.* (b) *Where the c_j's are integers, the vertices of $P[(33)]$ are integer-valued.*

Theorem (34a) generalizes to the following.

(35)   **Theorem.**   *For any two integral polymatroids P_1 and P_2 in $\mathbb{R}_E^+$, the vertices of $P_1 \cap P_2$ are integer-valued.*

The following technique for proving theorems like (34) is due to Alan Hoffman [7].

(36)   **Theorem.**   *The matrix $[a_{ij}]$ of the Hitchcock problem is totally unimodular — that is, the determinant of every square submatrix has value 0, 1, or −1.*

(37)   **Theorem.**   *Theorem (34) holds whenever $[a_{ij}]$ is totally unimodular.*

(38)   Let each V_p, $p = 1$ and 2, be a family of subsets of E such that any two members of P are either disjoint or else one is a subset of the other.

(39)   **Theorem.**   *The incidence matrix of the $V_1 \cup V_2$ of (38) is totally unimodular.*

Property (29) is a special case of (38). Property (38) is a special case of the following.

(40)   Let each V_p, $p = 1$ and 2, be a family of subsets of E such that for any $R \in V_p$ and $S \in V_p$ either $R \cap S = \emptyset$ or $R \cap S \in V_p$.

The incidence matrix of the $V_1 \cup V_2$ of (40) is generally not totally unimodular. However,

(41)   **Theorem.**   *From the incidence matrix of each V_p of (40), once can obtain, by subtracting certain rows from others, the incidence matrix of a family of mutually disjoint subsets of E. Thus, in the same way, one can obtain from the incidence matrix of the $V_1 \cup V_2$ of (40), a matrix of the Hitchcock type.*

(42)   **Theorem.**   *For any polymatroid $P(E, f)$ and any $x \in P(E, f)$, if $x(A) = f(A)$ and $x(B) = f(B)$ then either $A \cap B = \emptyset$ or $x(A \cap B) = f(A \cap B)$.*

Theorems (42), (41), and (34a) imply (35).

(43) Assuming that each V_p of (38) contains the set E, $L_p = V_p \cup \{\emptyset\}$ is a particularly simple lattice. For any non-negative non-decreasing function $f(i) = b_i$, $i \in V_p$, let $f(\emptyset) = -f(E)$. Then f is a β_0-function on L_p.

(44) The only integer vectors in a matroid polyhedron P are the vectors of the independent sets of the matroid, and these vectors are all vertices of P. Thus, (35) implies:

(45) **Theorem.** *Where P_1 and P_2 are the polyhedra of any two matroids M_1 and M_2 on E, the vertices of $P_1 \cap P_2$ are precisely the vectors which are vertices of both P_1 and P_2 — namely, the incidence vectors of sets which are independent in both M_1 and M_2.*

Where P_1, P_2, and P_3 are the polyhedra of three matroids on E, polyhedron $P_1 \cap P_2 \cap P_3$ generally has many vertices besides those which are vertices of P_1, P_2, and P_3.

Let $c = [c_j]$, $j \in E$, be any numerical weighting of the elements of E. In view of (45), the problem:

(46) Find a set J, independent in both M_1 and M_2, that has maximum weight-sum, $\sum_{j \in J} c_j$, is equivalent to the *l.p.* problem:

(47) Find a vertex x of $P_1 \cap P_2$ that maximizes $c \cdot x$.

(48) Assuming there is a good algorithm for recognizing whether of not a set $J \subseteq E$ is independent in M_1 or in M_2, there is a good algorithm for problem (46). This seems remarkable in view of the apparent complexity of matroid polyhedra in other respects. For example, a good algorithm is not known for the problem:

(49) Given a matroid $M_1 = (E, F_1)$ and given an element $e \in E$, minimize $|D|$, $D \subseteq E$, where $e \in D \notin F_1$;

Or the problems:

(50) Given three matroids M_1, M_2, and M_3, on E, and given an objective vector $c \in \mathbb{R}_E$, maximize $c \cdot x$ where $x \in P_1 \cap P_2 \cap P_3$.

Or maximize $\sum_{j \in J} c_j$ where $J \in F_1 \cap F_2 \cap F_3$.

V

Where f_1 and f_2 are β-functions on L_E, the dual of the *l.p.*:

(51) Maximize $c \cdot x = \sum_{j \in E} c_j x_j$, where

(52) For every $j \in E$, $x_j \geq 0$; and for every $A \in K$, $x(A) \leq f_1(A)$ and $x(A) \leq f_2(A)$; is the *l.p.*:

(53)   Minimize $f \cdot y = \sum_{A \in K} [f_1(A)y_1(A) + f_2(A)y_2(A)]$

where

(54)   For every $A \in K$, $y_1(A) \geq 0$ and $y_2(A) \geq 0$; and for every $j \in E$,

$$\sum_{j \in A \in K} [y_1(A) + y_2(A)] \geq c_j.$$

Combining systems (52) and (54) we get,

$$\sum_{j \in E} x_j \left(\sum_{j \in A \in K} [y_1(A) + y_2(A)] - c_j \right) \qquad (55)$$
$$+ \sum_{A \in K} y_1(A)[f_1(A) - \sum_{j \in A} x_j]$$
$$+ \sum_{A \in K} y_2(A)[f_2(A) - \sum_{j \in A} x_j] \geq 0.$$

Expanding and cancelling we get

$$c \cdot x \leq f \cdot y \qquad (56)$$

for any x satisfying (52) and any $y = (y_1, y_2)$ satisfying (54).

(57)   Equality holds in (56) if and only if equality holds in (55).

The *l.p.* duality theorem says that

(58)   If there is an x^0, a vertex of $P[(52)]$, which maximizes $c \cdot x$, then there is a $y^0 = (y_1^0, y_2^0)$, a vertex of $P[(54)]$, such that

$$c \cdot x^0 = f \cdot y^0, \qquad (59)$$

and hence such that y^0 minimizes $f \cdot y$.

For the present problem obviously there is such an x^0.

The vertices of (54) are not generally all integer-valued when the c_j's are. However,

(60)   **Theorem.**   *If the c_j's are all integers, then, regardless of whether f_1 and f_2 are integral, there is an integer-valued solution $y^4 = (y_1^4, y_2^4)$ of (54) which minimizes $f \cdot y$.*

Let $y^3 = (y_1^3, y_2^3)$ be any solution of (54) which minimizes $f \cdot y$.

(61)   For every $j \in E$, and $p = 1, 2$ let $c_j^p = \sum_{j \in A \in K} y_p^3$

For each $p = 1, 2$ consider the problem,

(62)   Minimize $f_p \cdot y_p = \sum_{A \in K} f_p(A) y_p(A)$ where

(63)   for every $A \in K$, $y_p(A) \geq 0$; and for every $j \in E$,

$$\sum_{j \in A \in K} y_p(A) \geq c_j^p.$$

(64)   By (21) for each p, there is an optimum solution, say y_p^4, to (62) having the following form:

(65)   The sets $A \in K$, such that $y_p^4(A) > 0$, form a nested sequence,

$$A_1 \subset A_2 \subset A_3 \subset \dots .$$

Since y_p^3 is a solution of (63), we have $f_p y_p^4 \leq f_p y_p^3$, for each p, and thus $f \cdot y^4 \leq f \cdot y^3$. Since $c_j^1 + c_j^2 \geq c_j$ for every $j \in E$, y^4 is a solution of (54), and hence y^4 is an optimum solution of (54). Thus, we have that

(66)   **Theorem.**   *There exists a solution y^4 of (54) which minimizes $f \cdot y$ and which has property (65) for each $p = 1, 2$.*

The problem, minimize $f \cdot y$ subject to (54) and also subject to $y_p(A) = 0$ for every $y_p^4(A) = 0$, has the form [(32), (33)] where $[a_{ij}]$ is the incidence matrix of a $V_1 \cup V_2$ as in (38). Thus, by (39) and (37), we have:

(67)   **Theorem.**   *If the c_j's are all integers then the y^4 of (66) can be taken to be integer-valued.*

In particular this proves (60).

An immediate consequence of (35), (60), and the *l.p.* duality theorem is

(68)   **Theorem.** $\max c \cdot x = \min f \cdot y$ *where $x \in P[(52)]$ and $y \in P[(54)]$.*
    If f is integral, x can be integral.
    If c is integral, y can be integral.

In particular, where f_1 and f_2 are the rank functions, r_1 and r_2, of any two matroids, $M_1 = (E, F_1)$ and $M_2 = (E, F_2)$, and where every $c_j = 1$, (68) implies:

(69)   **Theorem.** $\max |J| = \min[r_1(S) + r_2(E - S)]$, *where $J \in F_1 \cap F_2$, and where $S \subseteq E$.*

(A related result is given by Tutte [16]).

VI

(70)   **Theorem.**   *For each $i \in E'$, let Q_i be a subset of E. For each $A' \subseteq E'$, let $u(A') = \bigcup_{i \in A'} Q_i$. Let f be any integral β-function on L_E.*

Then $f'(A') = f(u(A'))$ is an integral β-function on $L_{E'} = \{A' : A' \subseteq E'\}$.

(71)   This follows from the relations

$$u(A' \cup B') = u(A') \cup u(B') \quad \text{and}$$
$$u(A' \cap B') \subseteq u(A') \cap u(B').$$

(72)   Applying (15) to f' yields a matroid on E'.

(73)   In particular, taking f to mean cardinality, if we let $J' \subseteq E'$ be a member of F' iff $|A'| \leq |u(A')|$ for every $A' \subseteq J'$, then $M' = (E', F')$ is a matroid.

(74)   Hall's SDR theorem says that: $|A'| \leq |u(A')|$ for every $A' \subseteq J'$ iff the family $\{Q_j\}$, $i \in J'$, has a system of distinct representatives, i.e., a transversal. A transversal of a family $\{Q_i\}$, $i \in J'$ is a set $\{j_i\}$, $i \in J'$, of distinct elements such that $j_i \in Q_i$. Thus,

(75)   **Theorem.**   *For any finite family $\{Q_i\}$, $i \in E'$, of subsets of E, the sets $J' \subseteq E'$ such that $\{Q_i\}$, $i \in J'$, has a transversal are the independent sets of a matroid on E' (called a transversal matroid).*

There are a number of interesting ways to derive (75). Some others are in [2], [3], [5], and [12]. The present derivation is the way (75) was first obtained and communicated.

The following is the same result with the roles of elements and sets interchanged.

(76)   *Let $J \in F_0$ iff, for some $J' \subseteq E'$, J is a transversal of $\{Q_i\}$, $i \in J'$. That is, let $J \in F_0$ iff J is a partial transversal of $\{Q_i\}$, $i \in E'$. Then $M_0 = (E, F_0)$ is a matroid.*

(77)   Thus, where P_0 is the polyhedron of M_0 and where P is the polyhedron of any other matroid, $M = (E, F)$, on E, the vertices of $P_0 \cap P$ are the incidence vectors of the M-independent partial transversals of $\{Q_i\}$, $i \in E'$.

By (8), the rank function r_0 of M_0 is, for each $A \subseteq E$,

(78) $r_0(A) = \min \left[|A_0| + |\{i : (A - A_0) \cap Q_i \neq \emptyset\}| \right]$
where $A_0 \subseteq A$.

Combining (69) and (78), we get

(79) $\max |J| = \min \left[r(A_1) + |A_0| + |E'| - |\{i : Q_i \subseteq A_1 \cup A_0\}| \right]$
    $= \min \left[r(u(A')) + |E'| - |A'| \right]$, where $J \in F_0 \cap F$, $A_0 \cup A_1 \subseteq E$, $A_0 \cap A_1 = \emptyset$, and $A' \subseteq E'$.

In particular, (79) implies the following theorem of Rado [14], given in 1942.

(80)   *For any matroid M on E, a family $\{Q_i\}$, $i \in E'$, of subsets of E, has a transversal which is independent in M iff $|A'| \leq r(u(A'))$ for every $A' \subseteq E'$.*

Taking the f of (70) to be r, (70), (15), and (80) imply:

(81) **Theorem.**   *For any matroid M on E, and any family $\{Q_i\}$, $i \in E'$, of subsets of E, the sets $J' \subseteq E'$ such that $\{Q_i\}$, $i \in J'$, has an M-independent transversal are the independent sets of a matroid on E'.*

(82)   A *bipartite graph* G consists of two disjoint finite sets, V_1 and V_2, of nodes and a finite set $E(G)$ of edges such that each member of $E(G)$ meets one node in V_1 and one node in V_2.

The following theorem of König is a prototype of (69).

(83) **Theorem.**   *For any bipartite graph G, $\max |J|$, $J \subseteq E(G)$, such that*

(a) *no two members of J meet the same node in V_1, and*

(b) *no two members of J meet the same node in V_2,*

equals $\min(|T_1| + |T_2|)$, $T_1 \subseteq V_1$, and $T_2 \subseteq V_2$, such that every member of $E(G)$ meets a node in T_1 or a node in T_2.

(84)   To get the Hall theorem, (74), from (83), let V_1 be the E' of (70), let V_2 be the E of (70), and let there be an edge in $E(G)$ which meets $i \in V_1$ and $j \in V_2$ iff $j \in Q_i$.

Clearly, if the family $\{Q_i\}$, $i \in E'$, has no transversal then, in (83), $\max |J| < |V_1|$. If the latter holds, then by (83), the T_1 of $\min(|T_1| + |T_2|)$, in (83), is such that

$$|V_1 - T_1| > |u(V_1 - T_1)|.$$

(85)   For the König-theorem instance, (83) of (69), the matroids $M_1 = (E, F_1)$ and $M_2 = (E, F_2)$ are particularly simple: Let $E = E(G)$. For $p = 1$ and $p = 2$, let $J \subseteq E(G)$ be a member of F_p iff no two members of J meet the same node in V_p.

(86)   Where P_1 and P_2 are the polyhedra of these two matroids, finding a vertex x of $P_1 \cap P_2$ which maximizes $c \cdot x$ is essentially the *optimal assignment problem*. That is, the Hitchcock problem where every $b_i = 1$.

(87)   Clearly, the inequality $x(A) \le r_p(A)$ is a face of P_p, that is, A is r_p-closed and r_p-inseparable, iff, for some node $v \in V_p$, A is the set of edges which meet v.

VII

(88)   Let $\{M_i\}$, $i \in I$, be a family of matroids, $M_i = (E, F_i)$, having rank functions r_i. Let $J \subseteq E$ be a member of F iff:

(89) $|A| \le \sum_i r_i(A)$ for every $A \subseteq J$.

Since $f(A) = \sum_i r_i(A)$ is a β-function on L_E,

(90) **Theorem.**   The $M = (E, F)$ of (88) *is a matroid, called the* sum *of the matroids* M_i.

In [5], and in [2], it is shown that

(91) **Theorem.** $J \subseteq E$ *satisfies* (89) *iff* J *can be partitioned into sets* J_i *such that* $J_i \in F_i$.

(92)   An algorithm, *MPAR*, is given there for either finding such a partition of J or else finding an $A \subseteq J$ which violates (89). That is, for recognizing whether or not $J \in F$.

(93)   The algorithm is a good one, assuming:

(94)   that a good algorithm is available for recognizing, for any $K \subseteq E$ and for each $i \in I$, whether or not $K \in F_i$.

(95)   The definition of a matroid $M = (E, F)$ is essentially that, modulo the ease of recognizing, for any $J \subseteq E$, whether or not $J \in F$, one has what is perhaps the easiest imaginable algorithm for finding, in any $A \subseteq E$, a maximum cardinality subset J of A such that $J \in F$.

(96)   In particular, by virtue of (90), assuming (94), *MPAR* provides a good algorithm for finding a maximum cardinality set $J \subseteq E$ which is partitionable into sets $J_i \in F_i$.

(97)   Assuming (94), *MPAR* combined with (19) is a good algorithm for, given numbers c_j, $j \in E$, finding a set J which is partitionable into sets $J_i \in F_i$ and such that $\sum_{j \in J} c_j$ is maximum.

Where r is the rank function of matroid $M = (E, F)$, let

(98) $r^*(A) = |A| + r(E - A) - r(E)$ for every $A \subseteq E$.

Substituting $r(E) = |E| - r^*(E)$, and A for $E - A$, in (98), yields

(99) $r(A) = |A| + r^*(E - A) - r^*(E)$.

(100) It is easy to verify that r^* is the rank function of a matroid $M^* = (E, F^*)$, e.g., that r^* satisfies (9). M^* is called *the* dual *of* M. By (99), $M^{**} = M$.

(101) By (98), $|J| = r^*(J)$ iff $r(E - J) = r(E)$. Therefore, $J \in F^*$ iff $E - J$ contains an M-basis of E, i.e., a *basis* of M. Thus, it can be determined whether or not $J \in F^*$ by obtaining an M-basis of $E - J$ and observing whether or not its cardinality equals $r(E)$.

Where r is the rank function of a matroid $M = (E, F)$, and where n is a non-negative integer, let

(102) $r^{(n)}(A) = \min[n, r(A)]$ for every $A \subseteq E$.

(103) Clearly, $r^{(n)}$ is the rank function of a matroid $M^{(n)} = (E, F^{(n)})$, called the *n-truncation* of M, such that $J \in F^{(n)}$ iff $J \in F$ and $|J| \leq n$.

(104) For matroids $M_1 = (E, F_1)$ and $M_2 = (E, F_2)$, and any integer $n \leq r_2(E)$, by (103) and (101), there is a set $J \in F_1 \cap F_2$ such that $|J| = n$ iff E can be partitioned into a set $J_1 \in F_1$ and a set $J_2 \in F_2^{(n)*}$. Theorem (91) says this is possible iff $|A| \leq r_1(A) + r_2^{(n)*}(n)(A)$ for every $A \subseteq E$. Using (102) and (98), this implies (69).

(105) Using *MPAR*, a maximum cardinality $J \in F_1 \cap F_2$ can be found as follows: Find a maximum cardinality set $H = J_1 \cup J_2$ such that $J_1 \in F_1$ and $J_2 \in F_2^*$. Extend J_2 to B, an M_2^*-basis of H. Clearly, B is an M_2^*-basis of E, and so $H - B \in F_1 \cap F_2$. It is easy to verify that $|H - B| = \max |J|$, $J \in F_1 \cap F_2$.

(106) It is more practical to go in the other direction, obtaining for a given family of matroids $M_i = (E, F_i)$, $i \in I$, an "optimum" family of mutually disjoint sets $J_i \in F_i$, by using the "matroid intersection algorithm" of (48) on the following two matroids $M_1 = (E_I, F_1)$ and $M_2 = (E_I, F_2)$. Let E_I consist of all pairs (j, i), $j \in E$ and $i \in I$. There is a 1–1 correspondence between sets $J \in E_I$ and families $\{J_i\}$, $i \in I$, of sets $J_i \subseteq E$, where J corresponds to the family $\{J_i\}$ such that $j \in J_i \iff (j, i) \in J$. Let $M_1 = (E_I, F_1)$ be the matroid such that $J \subseteq E_I$ is a member of F_1 iff the corresponding sets J_i are mutually disjoint — that is, if and only if the j's of the members of J are distinct. Let $M_2 = (E_I, F_2)$ be the matroid such that $J \subseteq E_I$ is a member of F_2 iff the corresponding sets J_i are such that $J_i \in F_i$.

(Nash-Williams has developed the present subject in another interesting way [13].)

VIII

(107) If $f(a)$ is a β-function on L and k is a not-too-large constant, then $f(a) - k$ is a β_0-function on L. It is useful to apply (15) to, non-β, β_0-functions.

(108) For example, let G be a graph having edge-set $E = E(G)$ and node-set $V = V(G)$. For each $j \in E$, let Q_j be the set of nodes which j meets. For every $A \subseteq E$, let $f(A) = |u(A)| - 1$. Then, by (70), $f(A)$ is a β_0-function on L_E.

(109) Applying (15) to this f yields a matroid, $M(G) = (E, F(G))$.

(110) The minimal dependent sets of a matroid $M = (E, F)$, i.e., the minimal subsets of E which are not members of F, are called the *circuits* of M.

(111) The circuits of $M(G)$ are the minimal non-empty sets $A \subseteq E$ such that $|A| = |u(A)|$.

(112) A set $J \subseteq E$ is a member of $F(G)$ iff J together with the set $u(J)$ of nodes is a forest in G.

IX

(113) Let G be a directed graph. For any $R \subseteq V(G)$, a *branching* B of G *rooted at R*, is a forest of G such that, for every $v \in V(G)$, there is a unique directed path in B (possibly having zero edges) from some node in R to v.

(114) The following problem is solved using matroid intersection

(115) Given any directed graph G, given a numerical weight c_j for each $j \in E = E(G)$, and given sets $R_i \subseteq V(G)$, $i \in I$, find edge-disjoint branchings B_i, $i \in I$, rooted respectively at R_i, which minimize $s = \sum_j c_j$, $j \in \bigcup_{i \in I} B_i$.

(116) The problem easily reduces to the case where each R_i consists of the same single node, $v_0 \in V(G)$. That is, find $n = |I|$ edge-disjoint branchings B_i, each rooted at node v_0, which minimize s.

(117) Where $F(G)$ is as defined in (109), let $J \subseteq E$ be a member of F_1 iff it is the union of n members of $F(G)$. By (91), $M_1 = (E, F_1)$ is a matroid.

(118) Let $J \subseteq E$ be a member of F_2 iff no more than n edges of J are directed toward the same node in $V(G)$ and no edge of J is directed toward v_0. Clearly, $M_2 = (E, F_2)$ is a matroid:

(119) **Theorem.**   *A set $J \subseteq E$ is the edge-set of n edge-disjoint branchings of G, rooted at node $v_0 \in V(G)$, iff $|J| = n(|V(G)| - 1)$ and $J \in F_1 \cap F_2$.*

This is a consequence of the following.

(120) **Theorem.**   *The maximum number of edge-disjoint branchings of G, rooted at v_0, equals the minimum over all C, $v_0 \in C \subset V(G)$, of the number of edges having their tails in C and their heads not in C.*

(121) There is an algorithm for finding such a family of branchings in G, and in particular for partitioning a set J as described in (119) into branchings as described in (119).

(122) Let P_1 and P_2 he the polyhedra of matroids M_1 and M_2 respectively. Let $H = \{x : x(E) = n(|V(G)| - 1)\}$.

It follows from (45) that

(123) A vector $x \in \mathbb{R}_E$ is a vertex of $P_1 \cap P_2 \cap H$ iff it is the incidence vector of a set J as described in (119).

(124) A variant of the matroid-intersection algorithm will find such an x which minimizes $c \cdot x$. The case $n = 1$ is treated in [4].

X

(125) Let each L_i be a commutative semigroup. We say $a \leq b$, for $\{a, b\} \subseteq L_i$, iff $a + d = b$ for some $d \in L_i$.

(126) A function f from L_0 into L_1 is called a ψ-function iff

(127) for every $\{a, d\} \subseteq L_0$, $f(a) \leq f(a + d)$; and

(128) for every $\{a, b, c\} \subseteq L_0$, $f(a + b + c) + f(c) \leq f(a + c) + f(b + c)$.

(129) L_i is called a ψ-semigroup iff, for $\{a, b, c\} \subseteq L_i$,

$$a + c + c = b + c + c \implies a + c = b + c.$$

For example, L_i is a ψ-semigroup if it is cancellative or if it is idempotent.

(130) **Theorem.**   If $f(\cdot)$ is a ψ-function from L_0 into L_1, $g(\cdot)$ is a ψ-function from L_1 into L_2, and L_1 is a ψ-semigroup, then $g(f(\cdot))$ is a ψ-function from L_0 into L_2.

(131) **Theorem.**   A function f from a lattice, L_0, into the non-negative reals, L_1, satisfies (128), where "+" in L_0 means "$\vee$" and "+" in L_1 means ordinary addition, iff f is non-decreasing, i.e., satisfies (127), and f is submodular.

(132) Thus, β-functions can be obtained by composing ψ-functions.

(133) **Theorem.**   A function f from the non-negative reals into the non-negative reals is a ψ-function, relative to addition in the image and preimage, iff it is non-decreasing and concave.

(134) **Theorem.**   A function f from a lattice, L_0, into a lattice, L_1, is a ψ-function, relative to joins "$\vee$" in each, iff it is a join-homomorphism, i.e., for every $\{a, b\} \subseteq L_0$, $f(a \vee b) = f(a) \vee f(b)$.

Let $h(S)$ be any real (integer)-valued function of the elements $S \in L$ of a finite lattice L. In principle, an (integral) non-decreasing submodular function f on L can be obtained recursively from h as follows:

(135) **Theorem.**   For each $S \in L$, let $g(S) = \min[h(S), g(A) + g(B) - g(A \wedge B)]$ where $A < S$, $B < S$, and $A \vee B = S$. Then g is submodular. For each $S \in L$, let $f(S) = \min g(A)$ where $S \leq A \in L$. Then f is submodular and non-decreasing. If h is submodular then $g = h$. If h is submodular and non-decreasing then $f = h$.

(A similar construction was communicated to me by D. A. Higgs.)

(136) The β-functions on a finite lattice L correspond to the members of a polyhedral cone $\beta(L)$ in the space of vectors $y = [y_A]$, $A \in L - \{\emptyset\}$. Where $y_\emptyset = 0$, $\beta(L)$ is the set of solutions to the system:

(137) $y_A + y_B - y_{A \vee B} - y_{A \wedge B} \geq 0$ and $y_{A \vee B} - y_A \geq 0$ for every $A \in L$ and $B \in L$.

(138) Characterizing the extreme rays of $\beta(L)$, in particular for $L = \{A : A \subseteq E\}$, appears to be difficult.

References

1. Dilworth, R.P., Dependence Relations in a Semimodular Lattice, *Duke Math. J.*, 11 (1944), 575–587.
2. Edmonds, J. and Fulkerson, D.R., Transversals and Matroid Partition, *J. Res. Nat. Bur. Standards*, 69B (1965), 147–153.
3. Edmonds, J., Systems of Distinct Representatives and Linear Algebra, *J. Res. Nat. Bur. Standards*, 71B (1967), 241–245.
4. Edmonds, J., Optimum Branchings, *J. Res. Nat. Bur. Standards*, 71B (1967), 233–240, reprinted with [5], 346–361.
5. Edmonds, J., Matroid Partition, *Math. of the Decision Sciences, Amer. Math Soc. Lectures in Appl. Math.*, 11 (1968), 335–345.
6. Gale, D., Optimal assignments in an ordered set: an application of matroid theory, *J. Combin. Theory*, 4 (1968) 176–180.
7. Hoffman, A.J., Some Recent Applications of the Theory of Linear Inequalities to Extremal Combinatorial Analysis, *Proc. Amer. Math. Soc. Symp. on Appl. Math.*, 10 (1960), 113–127.
8. Ingleton, A.W., A Note on Independence Functions and Rank, *J. London Math. Soc.*, 34 (1959), 49–56.
9. Kruskal, J.B., On the shortest spanning subtree of a graph, *Proc. Amer. Math. Soc.*, 7 (1956), 48–50.
10. Kuhn, H.W. and Tucker, A.W., eds., *Linear inequalities and related systems, Annals of Math. Studies*, no. 38, Princeton Univ. Press, 1956.
11. Lehman, A., A Solution of the Shannon Switching Game, *J. Soc. Indust. Appl. Math.*, 12 (1964) 687–725.
12. Mirsky, L. and Perfect, H., Applications of the Notion of Independence to Problems in Combinatorial Analysis, *J. Combin. Theory*, 2 (1967), 327–357.
13. Nash-Williams, C.St.J.A., An application of matroids to graph theory, *Proc. Int'l. Symposium on the Theory of Graphs*, Rome 1966, Dunod.
14. Rado, R., A theorem on Independence Relations, *Quart. J. Math.*, 13 (1942), 83–89.
15. Rado, R., A Note on Independence Functions, *Proc. London Math. Soc.*, 7 (1957), 300–320.
16. Tutte, W.T., Menger's Theorem for Matroids, *J. Res. Nat. Bur. Standards*, 69B (1965), 49–53.

Matching: A Well-Solved Class of Integer Linear Programs

Jack Edmonds[1] and Ellis L. Johnson[2]

[1] National Bureau of Standards, Washington, D.C., U.S.A.
[2] I.B.M. Research Center, Yorktown Heights, NY, U.S.A.

A main purpose of this work is to give a good algorithm for a certain well-described class of integer linear programming problems, called *matching problems* (or the *matching problem*). Methods developed for simple matching [2,3], a special case to which these problems can be reduced [4], are applied directly to the larger class. In the process, we derive a description of a system of linear inequalities whose polyhedron is the convex hull of the admissible solution vectors to the given matching problem. At the same time, various combinatorial results about matchings are derived and discussed in terms of graphs.

The *general integer linear programming problem* can be stated as:

(1)   Minimize $z = \sum_{j \in E} c_j x_j$, where c_j is a given real number, subject to

(2) x_j an integer for each $j \in E$;

(3) $0 \leq x_j \leq \alpha_j$, $j \in E$, where α_j is a given positive integer or $+\infty$;

(4) $\sum_{j \in E} a_{ij} x_j = b_i$, $i \in V$, where a_{ij} and b_i are given integers;

   V and E are index sets having cardinalities $|V|$ and $|E|$.

(5)   The integer program (1) is called a *matching problem* whenever

$$\sum_{i \in V} |a_{ij}| \leq 2$$

   holds for all $j \in E$.

(6)   A *solution* to the integer program (1) is a vector $[x_j]$, $j \in E$, satisfying (2), (3), and (4), and an *optimum solution* is a solution which minimizes z among all solutions. When the integer program is a matching problem, a solution is called a *matching* and an optimum solution is an *optimum matching*.

If the integer restriction (2) is omitted, the problem becomes a linear program. An optimum solution to that linear program will typically have fractional values. There is an important class of linear programs, called transportation or network flow problems, which have the property that for any integer right-hand side b_i, $i \in V$, and any cost vector c_j, $j \in E$, there is an optimum solution which has all integer x_j, $j \in E$. The class of matching probems includes that class of linear programs, but, in addition, includes problems for which omitting

M. Jünger et al. (Eds.): Combinatorial Optimization (Edmonds Festschrift), LNCS 2570, pp. 27–30, 2003.
© Springer-Verlag Berlin Heidelberg 2003

the integer restriction (2) results in a linear program with no optimum solution which is all integer.

Many interesting and practical combinatorial problems can be formulated as integer linear programs. However, limitations in the known methods for treating general integer linear programs have made such formulations of limited value. By contrast with general integer linear programming, the matching problem is *well-solved.*

(7)   **Theorem.**   *There is an algorithm for the general matching problem such that an upper bound on the amount of work which it requires for any input is on the order of the product of (8), (9), (10), and (11). An upper bound on the memory required is on the order of (8) times (11) plus (9) times (11).*

(8) $|V|^2$, the number of nodes squared;

(9) $|E|$, the number of edges;

(10) $\sum_{i \in V} |b_i| + 2 \sum_{\alpha_j < \infty} \alpha_j$;

(11) $\log(|V| \max |b_i| + |E| \max_{\alpha_j < \infty} \alpha_j) + \log(\sum_{j \in E} |c_j|)$.

(12)   **Theorem.**   *For any matching problem, (1), the convex hull P of the matchings, i.e., of the solutions to [(2), (3), and (4)], is the polyhedron of solutions to the linear constraints (3) and (4) together with additional inequalities:*

(13) $\sum_{j \in W} x_j - \sum_{j \in U} x_j \geq 1 - \sum_{j \in U} \alpha_j$.

There is an inequality (13) for every pair (T, U) where T is a subset of V and U is a subset of E such that

(14) $\sum_{i \in T} |a_{ij}| = 1$ for each $j \in U$;

(15) $\sum_{i \in T} b_i + \sum_{j \in U} \alpha_j$ is an odd integer.

The W in (13) is given by

(16) $W = \{j \in E : \sum_{i \in T} |a_{ij}| = 1\} - U$.

(17)   Let Q denote the set of pairs (T, U). The inequalities (13), one for each $(T, U) \in Q$, are called the *blossom inequalities* of the matching problem (1).

By Theorem (12), the matching problem is the linear program:

(18)   Minimize $z = \sum_{j \in E} c_j x_j$ subject to (3), (4), and (13).

(19)   **Theorem.**   *If c_j is an integral multiple of $\sum_{i \in V} |a_{ij}|$, and if the l.p. dual of (18) has an optimum solution, then it has an optimum solution which is integer-valued.*

Using *l.p.* duality, theorems (12) and (19) yield a variety of combinatorial existence and optimality theorems.

To treat matching more graphically, we use what we will call *bidirected* graphs. All of our graphs are bidirected, so *graph* is used to mean bidirected graph.

(20) A *graph* G consists of a set $V = V(G)$ of *nodes* and a set $E = E(G)$ of *edges*. Each edge has one or two *ends* and each end *meets* one node. Each end of an edge is either a *head* or a *tail*.

(21) If an edge has two ends which meet the same node it is called a *loop*. If it has two ends which meet different nodes, it is called a *link*. If it has only one end it is called a *lobe*. An edge is called *directed* if it has one head and one tail. Otherwise it is called *undirected, all-head* or *all-tail* accordingly.

(22) The node-edge incidence matrix of a graph is a matrix $A = [a_{ij}]$ with a row for each node $i \in V$ and a column for each edge $j \in E$, such that $a_{ij} = +2$, $+1$, 0, -1, or -2, according to whether edge j has two tails, one tail, no end, one head, or two heads meeting node i. (Directed loops are not needed for the matching problem.)

(23) If we interpret the capacity α_j to mean that α_j copies of edge j are present in graph G^α, then x_j copies of j for each $j \in E$, where $x = [x_j]$ is a solution of $[(2), (3), (4)]$, gives a subgraph G^x of G^α. The *degree* of node i in G^x is b_i, the number of tails of G^x which meet i minus the number of heads of G^x which meet i. Thus, where x is an optimum matching, G^x can be regarded as an "optimum degree-constrained subgraph" of G^α, where the b_i's are the degree constraints.

(24) A Fortran code of the algorithm is available from either author. It was written in large part by Scott C. Lockhart, who also wrote many comments interspersed through the deck to make it understandable. Several random problem generators are included.

It has been run on a variety of problems on a Univac 1108, IBM 7094, and IBM 360. On the latter, problems of 300 nodes, 1500 edges, $b = 1$ or 2, $\alpha = 1$, and randon c_j's from 1 to 10, take about 30 seconds. Running times fit rather closely a formula which is an order of magnitude better than our theoretical upper bound.

References

1. Berge, C., *Théorie des graphes et ses applications*, Dunod, Paris, 1958.
2. Edmonds, J., Paths, trees, and flowers, *Canad. J. Math.* 17 (1965), 449–467.
3. Edmonds, J., Maximum matching and a polyhedron with 0,1-vertices, *J. Res. Nat. Bur. Standards* 69B (1965), 125–130.
4. Edmonds, J., An introduction to matching, preprinted lectures, Univ. of Mich. Summer Engineering Conf. 1967.
5. Johnson, E.L., Programming in networks and graphs, Operation Research Center Report 65-1, Etchavary Hall, Univ. of Calif., Berkeley.
6. Tutte, W.T., The factorization of linear graphs, *J. London Math. Soc.* 22 (1947), 107–111.
7. Tutte, W.T., The factors of graphs, *Canad. J. Math.* 4 (1952), 314–328.
8. Witzgall, C. and Zahn, C.T. Jr., Modification of Edmonds' matching algorithm, *J. Res. Nat. Bur. Standards* 69B (1965), 91–98.

9. White, L.J., A parametric study of matchings, Ph.D. Thesis, Dept. of Elec. Engineering, Univ. of Mich., 1967.

10. Balinski, M., A labelling method for matching, Combinatorics Conference, Univ. of North Carolina, 1967.

11. Balinski, M., Establishing the matching poloytope, preprint, City Univ. of New York, 1969.

Theoretical Improvements in Algorithmic Efficiency for Network Flow Problems

Jack Edmonds[1] and Richard M. Karp[2]

[1] National Bureau of Standards, Washington, D.C., U.S.A.
[2] University of California, Berkeley, CA, U.S.A., formerly at I.B.M. Thomas
J. Watson Research Center

This paper presents new algorithms for the maximum flow problem, the Hitchcock transportation problem and the general minimum-cost flow problem. Upper bounds on the number of steps in these algorithms are derived, and are shown to improve on the upper bounds of earlier algorithms.

1   The Maximum Flow Problem

A network N is a directed graph together with an assignment of nonnegative capacity $c(u, v)$ to each arc (u, v). A *flow* is an assignment of a real number $f(u, v)$ to each arc so that

(i) $0 \leq f(u, v) \leq c(u, v)$;

(ii) for any fixed u, $\sum_v f(u, v) = \sum_v f(v, u)$.

One arc (t, s) is distinguished, and a flow which maximizes $f(t, s)$ is called *maximum*. Let f^* denote the maximum value of $f(t, s)$.

Ford and Fulkerson [1] have given a labelling algorithm to compute a maximum flow by repeated flow changes along "flow-augmenting paths". They do not specify which flow-augmenting path to choose.

In Fig. 1, let M be any positive integer.

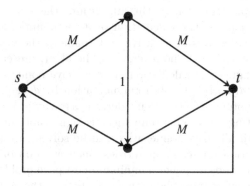

Fig. 1

M. Jünger et al. (Eds.): Combinatorial Optimization (Edmonds Festschrift), LNCS 2570, pp. 31–33, 2003.

Then, if a flow-augmenting path of length 3 is selected at each step, $2M$ augmentations will be needed to determine that $f^* = 2M$.

Let n denote the number of nodes of the network N.

Theorem 1 *If each flow augmentation is made along an augmenting path having a minimum number of arcs, then a maximum flow will be obtained after no more than $\frac{n^3-n}{4}$ augmentations.*

Let $\bar{c}$ be the average capacity of an arc (excluding the distinguished arc (t, s)).

Theorem 2 *If each flow augmentation is chosen to produce a maximum increase in $f(t, s)$, then the maximum flow will be obtained after no more than*

$$1 + \frac{n^2}{4}(1 + \frac{2}{n^2 - 2}) \, (2 \ln n + \ln \bar{c})$$

augmentations.

2   The Minimum-Cost Flow Problem

Assign to each arc (u, v) a nonnegative cost $d(u, v)$. Define the cost of a flow f as $\sum d(u, v)f(u, v)$. We seek a flow f of minimum cost subject to the constraint that $f(t, s) = f^*$.

Define the cost of a flow-augmenting path P as

$$\sum_{\substack{(u, v) \text{ a forward} \\ \text{arc in } P}} d(u, v) - \sum_{\substack{(u, v) \text{ a reverse} \\ \text{arc in } P}} d(u, v)$$

The following algorithm solves the minimum-cost flow problem: start with the zero flow, and repeatedly augment along a minimum-cost flow-augmenting path. Stop when a maximum flow is obtainted.

In a direct implementation of this algorithm, the selection of each flow-augmenting path requires the calculation of a minimum-cost path through a network which includes arcs of negative cost (namely, the reverse arcs of flow-augmenting paths). We show, however, that the algorithm can be modified so that, in each minimum-cost calculation, all arcs have nonnegative cost. This is done by replacing $d(u, v)$ at the kth augmentation by $d(u, v) + \pi^k(u) - \pi^k(v)$, where the "node potentials" $\pi^k(u)$ are derived as a by-product of finding the $(k - 1)$th augmenting path. This refinement is significant, since minimum-cost path problems with all costs nonnegative can be solved in $O(n^2)$ steps (where a step is an operation on scalar quantities, such as addition or comparison), whereas existing methods for general minimum-cost path problems require $O(n^3)$ steps. This refinement reduces the number of steps in the assignment problem, for example, from $O(n^4)$ to $O(n^3)$.

3   A Scaling Method for the Hitchcock Problem

An instance of the Hitchcock transportation problem is specified as follows:

$$\text{minimize} \sum_{i=1}^{n} \sum_{j=1}^{n} c_{ij} x_{ij}$$

subject to

$$\sum_{i} x_{ij} = b_j \qquad j = 1, 2, \ldots, n$$

$$\sum_{j} x_{ij} = a_i \qquad i = 1, 2, \ldots, m$$

$$x_{ij} \geq 0.$$

Here the "supplies" a_i and "demands" b_j are nonnegative integers auch that $\sum_{i=1}^{m} a_i = \sum_{j=1}^{n} b_j = B$.

A Hitchcock problem can be expressed as a minimum-cost flow problem on a network with $m + n + 2$ nodes, and can be solved within B flow augmentations. The scaling methods reduces this bound by applying the technique of flow augmentations to a sequence of approximate problems. In the pth approximate problem the ith supply is $\lfloor \frac{a_i}{2^p} \rfloor$, and the jth demand is $\lfloor \frac{b_j}{2^p} \rfloor$. A ficticious supply or demand is added in a standard way to establish a balance of supplies and demands. The original problem is the zeroth approximate problem. Approximate problems are solved successively, using the flow-augmentation path technique. A saving is effected by using, as a starting solution for approximate problem $p - 1$, twice the optimum solution for problem p. The over-all provess is shown to require only $n \log_2(1 + \frac{B}{n})$ flow augmentations.

(The material of this section was presented by the present authors under the title "A Technique for Accelerating the Solution of Transportation Problems" at the Second Annual Princeton Conference on Information Sciences and Systems, March, 1968.)

References

1. Ford, L.R. and Fulkerson, D.R. (1962) Flows in Networks, Princeton University Press, 1962.
2. Busacker, R.G. and Saaty, T.L. (1965) Finite Graphs and Networks, McGraw-Hill, 1965.

Connected Matchings

Kathie Cameron

Department of Mathematics
Wilfrid Laurier University
Waterloo, Ontario N2L 3C5
Canada
kcameron@wlu.ca

Abstract. A *connected matching* M in a graph G is a matching such that every pair of edges of M is joined by an edge of G. Plummer, Stiebitz and Toft introduced connected matchings in connection with their study of the famous Hadwiger Conjecture. In this paper, I prove that the connected matching problem is NP-complete for 0-1-weighted bipartite graphs, but polytime-solvable for chordal graphs and for graphs with no circuits of size 4.

1   Introduction

A *matching* is a set of edges, no two of which meet a common node. The optimum matching problem was well-solved by Edmonds [6,7]. Here I consider a particular type of matching. We say edges e and f are *joined by an edge* if an endpoint of e is joined to an endpoint of f. A *connected matching* M in a graph G is a matching such that every pair of edges of M is joined by an edge of G.

Plummer, Stiebitz and Toft [14] introduced connected matchings in connection with their study of the famous Hadwiger Conjecture, which says that the chromatic number of a graph G is at most the maximum number k for which G has a complete subgraph on k nodes, K_k, as a minor. In other words, any graph either has a $k - 1$ colouring of its nodes, or has K_k as a minor.

Their work suggests that connected matchings will play an important role in the solution of this conjecture: Note that by contracting the edges of a connected matching of size m, we obtain a complete subgraph of size m, and thus in graphs with no K_m minor, the maximum size of a connected matching is at most $m - 1$. In particular, the maximum size of a connected matching in a planar graph is at most 4.

Plummer, Stiebitz and Toft [14] showed that, for a given integer m, the problem of finding a connected matching of size at least m is NP-complete for general graphs. Independently, I showed the problem of finding a connected matching of weight at least m is NP-complete for 0-1-weighted bipartite graphs. The proof is in Section 2 below.

In this paper, I will give a polytime algorithm for finding a largest connected matching in chordal graphs. I will prove that in graphs with no circuit on four nodes, the maximum size of a connected matching is at most 5, and thus can be found in an obvious way.

M. Jünger et al. (Eds.): Combinatorial Optimization (Edmonds Festschrift), LNCS 2570, pp. 34–38, 2003.
© Springer-Verlag Berlin Heidelberg 2003

A notion closely related to that of connected matchings is induced matchings. An *induced matching* M in a graph G is a matching such that no two edges of M are joined by an edge of G; that is, an induced matching is a matching which forms an induced subgraph. Induced matchings have been studied by several authors. Stockmeyer and Vazirani [15] and I [1] independently proved that the problem of finding an induced matching of size at least m is NP-complete for bipartite graphs, and Ko and Shepherd [12] gave an easy proof that it is NP-complete for planar graphs. It has been shown that the problem of finding a largest induced matching can be solved in polytime in chordal graphs by me [1], in circular-arc graphs by Golumbic and Laskar [9], in cocomparability graphs by Golumbic and Lewenstein [10], in asteroidal-triple free graphs by Jou-Ming Chang [4] and me [2], in weakly chordal graphs by me, Sritharan, and Tang [3], and in Gavril's interval-filament graphs by me [2]. Interval-filament graphs include cocomparability graphs [8] and polygon-circle graphs [8], and polygon-circle graphs include chordal graphs [11], circular-arc graphs [11], circle graphs, and outer-planar graphs [13].

The *line-graph*, $L(G)$, of graph G has node-set $E(G)$, and an edge joining two nodes exactly when the edges of G they correspond to meet a common node. The *square*, G^2, of graph G has node-set $V(G)$, and two nodes are joined in G^2 exactly when they are joined by an edge or a path of two edges in G. A set of nodes is called *independent* if no two of them are joined by an edge. As pointed out in [1], for any graph G, every induced matching in G is an independent set of nodes in $[L(G)]^2$, and conversely.

A connected matching in G becomes a clique in $[L(G)]^2$. In [1], a set N of edges in a graph G was defined to be *neighbourly* if every pair of edges of N either meet a common node or are joined by an edge of G. For any graph G, neighbourly sets of edges correspond precisely to cliques in $[L(G)]^2$. Every connected matching is a neighbourly set, and a matching in a neighbourly set is a connected matching. Connected matchings in G correspond precisely to independent sets of nodes in $L(G)$ which are cliques in $[L(G)]^2$.

I will use uv to denote an edge with ends u and v.

2 The Connected Matching Problem Is NP-Complete for 0-1 Weighted Bipartite Graphs

Consider the problems

(C) Max clique: Given a positive integer m and an arbitrary graph G, does this graph have a clique of size at least m ?

and

(WBCM) Max connected matching in 0-1 weighted bipartite graphs: Given a positive integer m and a 0-1 weighted bipartite graph B, does B have a connected matching of weight at least m?

(WBCM) is clearly in NP. Here is a formulation of (C) as an instance of (WBCM). Since (C) is NP-complete, it follows that (WBCM) is NP-complete.

Given a graph G, construct a bipartite graph B as follows. For each node v of G, put two nodes v' and v'' in B, and join them by an edge with weight 1. For each edge uv of G, put edges $u'v''$ and $v'u''$ in B, and give these edges weight 0. If C is a clique of size m in G, $\{v'v'' : v \in C\}$ is a connected matching in B of weight m. Conversely, if M is a connected matching in B with weight m, then M contains m edges of the form $v'v''$. $\{v : v'v'' \in M\}$ is a clique in G of size m.

3   Finding Largest Connected Matchings in Chordal Graphs

A graph is called *chordal* if it has no chordless circuits on four or more nodes.

As mentioned in the introduction, connected matchings are precisely matchings in neighbourly sets. In [1] it is proved that neighbourly sets in chordal graphs are what I call *clique-neighbourhoods*: a clique together with edges meeting the nodes of the cliques. Thus, to find a largest connected matching in a chordal graph, we can find a largest matching contained in a clique-neighbourhood.

Note that for a clique-neighbourhood N where the clique C has k nodes, the size of a largest matching equals the size l of a largest matching in the bipartite graph of edges from nodes of C to nodes met by $N - C$ plus $\lfloor (k - l)/2 \rfloor$.

To find a largest matching contained in a clique-neighbourhood, we can restrict ourselves to clique-neighbourhoods where the clique is a maximal clique of the graph.

It follows from the fact that chordal graphs have a simplicial ordering [5] (an ordering such that the after-neighbours of a node form a clique) that chordal graphs have only a linear number of maximal cliques - they are of the form: a node together with its after-neighbours in the simplicial ordering. Thus given a chordal graph, enumerate all maximal cliques (using the simplicial ordering) and then for each, find the size of a largest matching from the clique nodes to other nodes. The largest connected matching is obtained when (the number of edges in the matching from clique nodes to other nodes) plus (the round-down of one half the number of nodes of the clique not met by the matching) is largest.

4   Connected Matchings in Graphs with No Circuits on Four Nodes

C_4 is the circuit with four nodes. We consider graphs which do not have C_4 as a subgraph. The Peterson graph is such a graph with a connected matching of size 5. The following theorem shows this is best possible.

Theorem 1. *The maximum size of a connected matching in a graph with no C_4 as a subgraph is 5.*

Proof. Let $G = (V, E)$ be a graph with no C_4 as a subgraph. Suppose G contains a connected matching M of size 6, say $M = \{s_i t_i : 1 \le i \le 6\}$.

There is an edge between every pair of edges of M. Without loss of generality, s_1 is joined to s_2, s_3, and s_4.

The following simple observations will be used below.

(1) If s_1 is joined to s_i and s_j, $i \neq j$, and $s_i t_j \in E$, then s_1, s_i, t_j, s_j would be a C_4.
(2) If s_1 is joined to s_i, s_j, and s_k, and $s_i s_j, s_j s_k \in E$, then s_1, s_i, s_j, s_k would be a C_4.

Suppose first that s_1 is also joined to s_5. By (1), there are no edges between $\{s_i : 2 \leq i \leq 5\}$ and $\{t_i : 2 \leq i \leq 5\}$ other than the M-edges. Without loss of generality, by (2), we can assume the only edges with both ends in $\{s_i : 2 \leq i \leq 5\}$ are possibly $s_2 s_3$ and $s_4 s_5$. Thus there are all possible edges with both ends in $\{t_i : 2 \leq i \leq 5\}$ except possibly $t_2 t_3$ and $t_4 t_5$. But then t_2, t_4, t_3, t_5 is a C_4.

So s_1 is not joined to s_5 or s_6, and thus, without loss of generality, t_1 is joined to t_5 and t_6.

By (1), the only edges between $\{s_i : 2 \leq i \leq 4\}$ and $\{t_i : 2 \leq i \leq 4\}$ are the M-edges, and analogously, the only edges between $\{s_i : 5 \leq i \leq 6\}$ and $\{t_i : 5 \leq i \leq 6\}$ are the M-edges. By (2), there is at most one edge with both ends in $\{s_i : 2 \leq i \leq 4\}$. Without loss of generality, we have either structure

(A) $s_2 s_4 \in E$ and $t_2 t_3$, $t_3 t_4 \in E$,
    or else
(B) $t_2 t_4, t_2 t_3$, $t_3 t_4 \in E$.

Edge $s_5 t_5$ is joined to each of the M-edges $s_i t_i$, $2 \leq i \leq 4$. It may be joined by an edge of the form $s_5 s_i$, $s_5 t_i$, or $t_5 t_i$; call these type 1, type 2, and type 3 edges respectively. An edge $t_5 s_i$ would create a C_4 : s_1, s_i, t_5, t_1.

If there are two type 1 edges, say $s_5 s_i$ and $s_5 s_j$, then s_1, s_i, s_5, s_j is a C_4.

Suppose there is both a type 1 edge, $s_5 s_i$, and a type 2 edge, $s_5 t_j$. If $t_i t_j \in E$, then s_5, s_i, t_i, t_j is a C_4. If $t_i t_j \notin E$, then we have structure (A), and without loss of generality, $i = 2$ and $j = 4$. Then s_2, s_4, t_4, s_5 is a C_4.

Suppose there are two type 2 edges, say $s_5 t_i$ and $s_5 t_j$. Then t_i, s_5, t_j, t_k is a C_4, where k is the one of 2, 3, 4 different from i and j unless we are have structure (A) and $\{i, j\} = \{2, 3\}$ or $\{i, j\} = \{3, 4\}$. So, without loss of generality, say $s_5 t_2$ and $s_5 t_3$ are type 2 edges, $s_5 t_4 \notin E$, and we have structure (A). If the edge joining $s_4 t_4$ and $s_5 t_5$ is a type 1 edge, $s_5 s_4$, then s_5, s_4, t_4, t_3 is a C_4. If the edge joining $s_4 t_4$ and $s_5 t_5$ is a type 3 edge, $t_5 t_4$, then t_3, t_4, t_5, s_5 is a C_4.

Thus there must be two type 3 edges, say $t_5 t_i$ and $t_5 t_j$. Then t_i, t_5, t_j, t_k is a C_4, where k is the one of 2, 3, 4 different from i and j unless we have structure (A) and $\{i, j\} = \{2, 3\}$ or $\{i, j\} = \{3, 4\}$. So, without loss of generality, say $t_5 t_2$ and $t_5 t_3$ are type 3 edges, $t_5 t_4 \notin E$, and we have structure (A). If the edge joining $s_4 t_4$ and $s_5 t_5$ is a type 2 edge, $s_5 t_4$, then t_3, t_4, s_5, t_5 is a C_4. The one remaining case is that the edge joining $s_4 t_4$ and $s_5 t_5$ is a type 1 edge, $s_5 s_4$. Edge $s_6 t_6$ must be joined to each of $s_2 t_2$, $s_3 t_3$, and $s_4 t_4$. Let's use the same notion of type for these edges as we did for the edges joining $s_5 t_5$ to $s_2 t_2$, $s_3 t_3$, and $s_4 t_4$. It must be that $s_6 t_6$ is joined to $s_2 t_2$, $s_3 t_3$, and $s_4 t_4$ by two type 3 edges and one type 1 edge since otherwise replacing 5 by 6 in the argument above, we are finished. In fact, it must be that either $t_6 t_2$ and $t_6 t_3$ are type 3 edges and $s_6 s_4$

is type 1 or t_6t_3 and t_6t_4 are type 3 edges and s_6s_2 is type 1. In the first case, t_2, t_5, t_3, t_6 is a C_4. In the second case, since s_5t_5 and s_6t_6 must be joined by an edge, it follows from the analogue of (1) obtained by interchanging s's and t's that either $s_5s_6 \in E$ or $t_5t_6 \in E$. In the first case, s_2, s_4, s_5, s_6 is a C_4. In the second case, t_3, t_4, t_6, t_5 is a C_4.

It follows that G does not contain a connected matching of size 6.

Acknowlegements. Research supported by the Natural Sciences and Engineering Research Council of Canada (NSERC) and Équipe Combinatoire, Université Pierre et Marie Curie (Paris IV), France.

References

1. Kathie Cameron, Induced matchings, *Discrete Applied Mathematics* **24** (1989), 97–102.
2. Kathie Cameron, Induced matchings in Intersection Graphs, 3-page abstract in: *Electronic Notes in Discrete Mathematics* **5** (2000); paper submitted for publication.
3. Kathie Cameron, R. Sritharan, and Yingwen Tang, accepted for publication in *Discrete Mathematics*.
4. Jou-Ming Chang, Induced matchings in asteroidal triple-free graphs, manuscript, April 2001.
5. G. A. Dirac, On rigid circuit graphs, *Abh. Math. Sem. Univ. Hamburg* **25** (1961), 71–76.
6. Jack Edmonds, Paths, trees, and flowers, *Canad. J. Math.* **17** (1965), 449–467.
7. Jack Edmonds, Maximum matching and a polyhedron with $0, 1$-vertices, *J. Res. Nat. Bur. Standards Sect. B* **69B** (1965), 125–130.
8. F. Gavril, Maximum weight independent sets and cliques in intersection graphs of filaments, *Information Processing Letters* **73** (2000), 181–188.
9. Martin Charles Golumbic and Renu C. Laskar, Irredundancy in circular-arc graphs, *Discrete Applied Mathematics* **44** (1993), 79–89.
10. Martin Charles Golumbic and Moshe Lewenstein, New results on induced matchings, *Discrete Applied Mathematics* **101** (2000), 157–165.
11. S. Janson and J. Kratochvil, Threshold functions for classes of intersection graphs, *Discrete Mathematics* **108** (1992), 307–326.
12. C. W. Ko and F.B. Shepherd, Adding an identity to a totally unimodular matrix, London School of Economics Operations Research Working Paper, LSEOR 94.14, July 1994.
13. Alexandr Kostochka and Jan Kratochvíl, Covering and coloring polygon-circle graphs, *Discrete Mathematics* **163** (1997), 299–305.
14. Michael D. Plummer, Michael Stiebitz and Bjarne Toft, On a special case of Hadwiger's conjecture, manuscript June 2001.
15. Larry J. Stockmeyer and Vijay V. Vazirani, NP-completeness of some generalizations of the maximum matching problem, *Information Processing Letters* **15** (1982), 14–19.

Hajós' Construction and Polytopes

Reinhardt Euler

Faculté des Sciences, 20 Avenue Le Gorgeu, 29285 Brest Cedex, France,
`Reinhardt.Euler@univ-brest.fr`

Abstract. Odd cycles are well known to induce facet-defining inequalities for the stable set polytope. In graph coloring odd cycles represent the class of 3-critical graphs. We study Hajós' construction to obtain a large class of n-critical graphs ($n > 3$), which properly generalize both cliques and odd cycles, and which again turn out to be facet-inducing for the associated stable set polytope.

1 Introduction

Given a finite graph $G = (V, E)$ and a set of colors $C = \{1, ..., n\}$ an *n-coloring* of G is a function $f : V \mapsto C$ such that $f(u) \neq f(v)$ for all edges uv of G. Every color class $f^{-1}\{i\}$ forms a *stable set*, i.e. a subset of vertices no two of which are joined by an edge. An n-coloring thus represents a partition of V (or $V(G)$) into n stable sets. The minimum number n for which G has an n-coloring is called the *chromatic number* $\chi(G)$ of G, and G is called $\chi(G)$-*chromatic*. A *homomorphism* of a graph G into a graph H is a mapping $\varphi : V(G) \mapsto V(H)$ such that $\varphi(x)\varphi(y)$ is an edge of H if xy is an edge of G. Therefore, an n-coloring of G may be considered as a homomorphism of G into the complete graph (or *clique*) K_n. Finally, we call a connected graph G *n-critical*, if $\chi(G) = n$ but $\chi(G \setminus e) = (n-1)$ for any edge e of G. This definition of criticality allows to establish a close relationship with a number of concepts from polyhedral theory such as "rank-criticality" (due to Chvàtal [3]) or "bipartite subgraphs" that might now be generalized to "n-colorable subgraphs" and whose associated polytopes are an interesting new subject in its own: we just mention at this place that an n-critical graph $G = (V, E)$ gives rise to an inequality of the form $\sum_{e \in E} x_e \leq |E| - 1$, which is valid for the (n-1)-colorable subgraph polytope and which can be lifted to a global facet-defining inequality. In this paper we will concentrate on the *stable set polytope P(G)* associated with certain graphs G, i.e. the convex hull of the incidence vectors of all stable sets in G. It will turn out that any n-critical graph presented gives rise to a facet-defining inequality of the induced stable set polytope, a result indicating that homomorphisms might be an interesting subject to study within polyhedral combinatorics.

Let us return to n-critical graphs and recall that there is just one 2-critical graph, K_2, and that odd cycles are well known to constitute the class of 3-critical graphs. Few is known for n exceeding 3, and it is one of our aims to describe a large class of n-critical graphs for $n > 3$. The basic tool to be used is *Hajós' construction* (cf. [1], [4]): call a graph *Hajós-n-constructible* if it can be obtained

M. Jünger et al. (Eds.): Combinatorial Optimization (Edmonds Festschrift), LNCS 2570, pp. 39–47, 2003.
© Springer-Verlag Berlin Heidelberg 2003

from complete graphs K_n by repeated application of the following two operations (see Figure 1 for an illustration):

(a) Let G_1 and G_2 be already obtained disjoint graphs with edges x_1y_1 and x_2y_2. Remove x_1y_1 and x_2y_2, identify x_1 and x_2, and join y_1 and y_2 by a new edge.

(b) Identify independent vertices (i.e. apply a homomorphism).

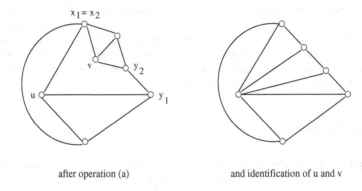

after operation (a)   and identification of u and v

Fig. 1. Hajós' construction for $G_1 = G_2 = K_4$

We call any such application (part (b) may be empty) a *Hajós step*. As shown by Hajós in 1961 this construction allows a characterization of n-chromatic graphs in the following way:

Theorem 1 (Hajós (1961)) *A graph has chromatic number at least n if and only if it contains a Hajós-n-constructible subgraph. Every n-critical graph is Hajós-n-constructible.*

According to [4] no interesting applications of this theorem have been found so far; in particular, no *explicit* description of n-critical graphs is available for $n > 3$. (How about the *automatic* generation of all such graphs for small n?) In a recent paper [5] we have generalized the notion of odd cycles to that of $(K_n \setminus e)$-cycles, a new class of n-critical graphs that allowed us to fully characterize the 3-colorability of infinite planar Toeplitz graphs. These graphs have also been shown to be facet-inducing for the associated stable set polytope. In this paper we are going to present even more general classes of n-critical graphs: the idea is to repeatedly apply Hajós' construction to the initial graph K_n.

We will present our results in two parts: first we will study the case that part (b) in Hajós' construction is always empty. Afterwards we will allow part (b) as part of a Hajós step, but slightly refined in order to avoid redundancies. As already done with respect to $(K_n \setminus e)$-cycles, we will describe the consequences for the associated stable set polytopes.

2   Hajós' Construction without Homomorphisms

Let G_1 be an n-critical graph, $G_2 = K_n$ and x_1y_1, x_2y_2 be arbitrary edges of G_1 and G_2, respectively. If G^* is the graph resulting from one application of operation (a) we have the following:

Theorem 2 G^* *is again n-critical.*

Proof: a) Suppose G^* is (n-1)-colorable. Then y_1 and y_2 must be colored differently with respect to this (n-1)-coloring, and the same holds for x_1 and y_1 (because x_1 has to be colored as y_2) . But then G_1 would have an (n-1)-coloring, a contradiction. Since G^* has an n-coloring, $\chi(G^*) = n$. b) We have to show that $\chi(G^* \setminus e) = (n-1)$ for any edge e of G^*. b1) e is an edge of $G_1 \setminus x_1y_1$: then $G_1 \setminus e$ has an (n-1)-coloring in which y_1 is colored differently from x_1. Now extend this (n-1)-coloring to G^* by giving y_2 the color of x_1 and by coloring the remaining vertices with the (n-2) colors left. b2) $e = y_1y_2$: $G_1 \setminus x_1y_1$ has an (n-1)-coloring which can be extended to G^* by coloring y_2 as y_1 (or x_1) and the remaining (n-2) vertices as in b1). b3) e is an edge of $K_n \setminus x_2y_2$: then $K_n \setminus \{x_2y_2, e\}$ can be colored with (n-1) colors such that x_2 and y_2 receive different colors; this coloring can be patched with an (n-1)-coloring of $G_1 \setminus x_1y_1$ (giving the same color to x_1 and y_1) to produce an (n-1)-coloring of G^*. This terminates our proof.

Starting with $G_1 = K_n$, which is n-critical, we hereby obtain a first class of n-critical graphs $\mathcal{G} = \{G_1, G_2, ..., G_m, ...\}$. For $n = 3$ we don't get anything new: G_m is just an odd cycle of length 2m+1.

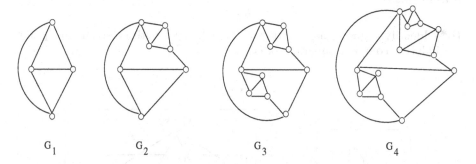

G_1 G_2 G_3 G_4

Fig. 2. A series of 4-critical graphs

If n=4, however (cf. Figure 2 for the first members of such a sequence), we obtain an interesting class of planar graphs: all of their faces are *odd*, i.e. delimited by odd cycles. They are of particular importance for a polyhedral study of the stable set problem in planar graphs.

For arbitrary n we get a class of graphs that properly generalize cliques and odd cycles. That they are again facet-inducing is shown by our next theorem:

Theorem 3 *Let* $G_m = (V_m, E_m)$ *be obtained from* $G_1 = K_n$ *by (m-1) applications of operation (a) as indicated above. Then the inequality* $x(V_m) := \sum_{v \in V_m} x_v \leq m$ *defines a facet of the induced stable set polytope* $P(G_m)$.

Proof: By Theorem 2, G_m is n-critical, i.e. $G_m \setminus e$ has an (n-1)-coloring for any edge $e \in E_m$. Since $|V_m| = (n-1)m+1$, there is a stable set S in $G_m \setminus e$ with $|S| = m+1$. The size of a largest stable set in G_m, however, is m. It follows that G_m is *rank-critical* and since it is also connected, Chvàtal's result [3] implies that the inequality $x(V_m) \leq m$ defines a facet of $P(G_m)$.

Theorem 3 leads to the following questions:

1) Is *any* planar graph with only odd faces facet-inducing for the associated stable set polytope?

2) Is there an efficient separation algorithm for this class of inequalities?

3   Hajós' Construction – The General Case

In this section we are going to refine operation (b) of Hajós' construction: the aim is to allow a non-redundant generation of n-critical graphs. Such redundancies may occur if, for instance, a vertex of $K_n \setminus \{x_2, y_2\}$ is identified with a non-neighbor of x_1, or if the vertices of $K_n \setminus \{x_2, y_2\}$ are identified with the complete G_1-neighborhood of x_1 (so that this vertex becomes superfluous). Our refinement is as follows:

(b*) Identify at most (n-3) vertices of $K_n \setminus \{x_2, y_2\}$ with neighbors of x_1 that induce a complete subgraph of G_1.

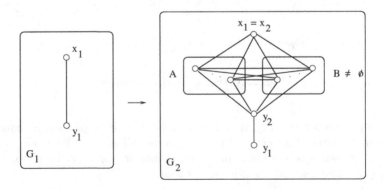

Fig. 3. Refinement of operation (b)

Figure 3 illustrates this refinement: $K_n \setminus \{x_2, y_2\}$ now splits up into an overlap with vertices of G_1, say A, and a set of new vertices, say B, which is supposed to be non-empty. We do not know whether *all* n-critical graphs can be constructed this way, but we certainly properly generalize the notion of $(K_n \setminus e)$-cycles as introduced in [5]: just observe that we get such a $(K_n \setminus e)$-cycle, if we start with an arbitrary edge in G_1 and then systematically choose x_1 to be y_2 and y_1 as before.

As in the previous section we can show

Theorem 4 *If G_1 is n-critical, then G_2 is n-critical, too.*

To prove this we proceed basically as in Theorem 2; the only difference occurs when the edge e of G_1 is incident with A: if e is not incident with x_1, then we can always extend the (n-1)-coloring of $G_1 \setminus e$ to G_2, and if e is incident with x_1, we color one new vertex with the color of y_1 (different from that of x_1) in order to have one color left for vertex y_2.

Again we can start with $H_1 = K_n$ to obtain a second class of n-critical graphs $\mathcal{H} = \{H_1, H_2, ..., H_m, ...\}$, and, as before, we don't get anything but odd cycles if n=3. Just note, that we may loose planarity if "overlaps" are allowed.

To come back to polyhedral aspects we may ask, whether and how these graphs H_m are facet-inducing for the associated stable set polytope. For this we first present a procedure that adjusts the coefficients of the linear inequality supposed to be facet-defining at each Hajós step:

Procedure "Facet-extension": Step 1: *Choose any edge $x_1 y_1$ in H_m.* **Step 2:**

Determine $w_0^{m+1} := w^m(H_m \setminus x_1 y_1)$, the maximum w^m-weight of a stable set in $H_m \setminus x_1 y_1$, and set $\delta_m := w_0^{m+1} - w_0^m$, the increase in weight caused by the deletion of $x_1 y_1$. **Step 3:** *For all new vertices $v \in (B \cup \{y_2\})$ set $w^{m+1}(v) := \delta_m$ and for all vertices $v \in A$ modify the current coefficient $w^m(v)$ to $w^{m+1}(v) := w^m(v) + \delta_m$.*

Figure 4 illustrates a series of 4-critical graphs together with the coefficients of their associated inequality.

It is not difficult to see that $1 \le \delta_m \le min(w^m(x_1), w^m(y_1))$. We are now going to show that the inequality $(w^m)^T x \le w_0^m$ defines a facet of $P(H_m)$ for all $m \in \mathbb{N}$. For this we need two (technical) results:

Lemma 1 *For every vertex v in H_m there is a stable set S of weight w_0^m containing v.*

For a proof we proceed by induction on m. The case $m = 1$ is clear; so let the assumption be true for H_{m-1} and consider the graph H_m obtained by an additional Hajós step (cf. Figure 3). For $v \in A$, by induction hypothesis, there is a stable set of weight w_0^{m-1} containing v, and by definition of $w^m(v)$ we are done. For $v = y_2$ observe that it can be added to the stable set of weight w_0^{m-1} containing x_1 and if v is one of the remaining new vertices one may add it to the corresponding stable set containing y_1. Finally, since B is nonempty, the assumption clearly holds for x_1 and y_1.

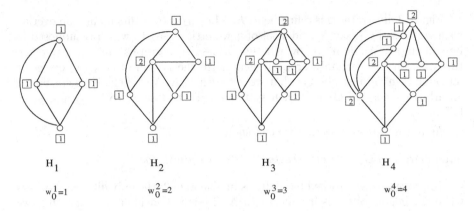

Fig. 4.

Lemma 2 *For every edge uv in H_m and clique-inducing neighborhood $N(u) \subseteq V_m$ with $|N(u)| \leq (n-3)$ there is a stable set S of weight w_0^m such that $S \cap (\{u,v\} \cup N(u)) = \emptyset$.*

Again, we proceed by induction on m. The case m=1 is obvious. Suppose that the statement is true for the graph H_{m-1}:

a) uv is an edge in H_{m-1} and $u \notin A, u \neq x_1$: by induction hypothesis there is a stable set S of weight w_0^{m-1} in H_{m-1} such that $S \cap (\{u,v\} \cup N(u)) = \emptyset$, in particular, $N(u) \subseteq V_{m-1}$. Now if $S \cap A \neq \emptyset$, then S is a w_0^m - weight stable set in H_m, and we are done. If, however, $S \cap A = \emptyset$, and $x_1 \in S$, then $S \cup y_2$ is a stable set of weight w_0^m in H_m having the desired property; if $x_1 \notin S$, there is a vertex $x \in B$ that we can add to S in a similar way.

b) uv is an edge in H_{m-1} and $u \in A$:

b1) $N(u) \cap B = \emptyset$: by induction hypothesis there is a w_0^{m-1} - weight stable set S in H_{m-1} such that $S \cap (\{u,v\} \cup N(u)) = \emptyset$. If $S \cap A \neq \emptyset$, S is the desired stable set. So let $S \cap A = \emptyset$. If $y_1 \in S$, add $x \in B$ to S, and we are done. If $y_1 \notin S$ it is sufficient to add y_2 to S.

b2) $N(u) \cap B \neq \emptyset$: then $N(u) \subseteq (\{x_1, y_2\} \cup A \cup B)$. Moreover, $x_1 \in N(u)$ implies $y_2 \notin N(u)$, and by induction hypothesis there is a w_0^{m-1} - weight stable set S in H_{m-1} such that $S \cap (\{x_1, y_1\} \cup A) = \emptyset$: now add y_2 to S and we are done. If finally, $x_1 \notin N(u)$, the w_0^m - weight stable set S of H_m containing both x_1 and y_1 has the desired property.

c) uv is an edge in H_{m-1} and $u = x_1$ (hence $v \in A$): by induction hypothesis there is a w_0^{m-1} - weight stable set S in H_{m-1} with $S \cap (\{u, y_1\} \cup N(u)) = \emptyset$. If $S \cap A \neq \emptyset$, S has the desired property. If however, $S \cap A = \emptyset$, because of $y_2 \notin N(u)$ we can add y_2 to S to obtain a w_0^m - weight stable set with the desired property.

d) uv is an edge in $H_m \setminus H_{m-1}$:

d1) $u = x_1$ and $v \in B$: by Lemma 1, we may suppose that $A \subseteq N(u)$ and by induction hypothesis there is a w_0^{m-1} - weight stable set S in H_{m-1} with $S \cap (\{u, y_1\} \cup A) = \emptyset$. Add y_2 to S and we are done. The converse situation: $v = x_1$ and $u \in B$ can be handled in a similar way.

d2) $u \in A$ and $v \in B \cup \{y_2\}$: again with $A \subseteq N(u)$ there is a vertex in $B \cup \{y_2\}$ and a w_0^{m-1} - weight stable set S in H_{m-1} with $S \cap (\{x_1, y_1\} \cup A) = \emptyset$; just add that vertex to S; the converse situation can be treated analogously.

d3) the case that both u and v are in $B \cup y_2$ is similar to case d2).

d4) $u = y_1$ and $v = y_2$: in H_{m-1} there is a w_0^{m-1} - weight stable set S with $S \cap (\{x_1, y_1\} \cup N(u)) = \emptyset$. But there is also a vertex $x \in B$ that we can add to S to obtain a stable set with the desired property. The converse case, i.e. $u = y_2$ and $d = y_1$, is similar to case d2). This completes the proof of Lemma 2.

We are now able to show

Theorem 5 *The inequality* $(w^m)^T x \leq w_0^m$ *defines a facet of* $P(H_m)$, *for all* $m \in I\!N$.

The proof is again by induction on m. Validity of our inequality is clear by construction, and for $m = 1$ the inequality $x(V_1) \leq 1$ is well known to be facet-defining. So suppose that $H_m = (V_m, E_m)$ is the n-critical graph obtained after $(m - 1)$ Hajós steps (with or without homomorphism) and that $(w^m)^T x \leq w_0^m$ is a facet-defining inequality for $P(H_m)$. It is sufficient to exhibit a linearly independent set of $|V_{m+1}|$ incidence vectors of stable sets all having weight w_0^{m+1}. Figure 5 illustrates the corresponding matrix M_{m+1}.

Observe that M_{m+1} is defined inductively: M_m is the submatrix surrounded by heavy lines. Also, the second last block of lines of M_{m+1} comes from the stable set S, whose existence is guaranteed by Lemma 2, and the last line corresponds to the stable set of weight w_0^{m+1} that is obtained by deleting the edge $x_1 y_1$ from H_m. It is not difficult to check that M_{m+1} has full rank, and this completes our proof.

4   Conclusion

We are not sure at this moment, whether we really grasp *all* of the n-critical graphs by our "general" construction. Figure 6 illustrates a class of 4-critical graphs, interesting for further study in this direction:

Again, any member of this class is facet-inducing for the associated stable set polytope: the corresponding inequality is

$$(k + 1)\left(\sum_{i=1}^{2k+1} x_i\right) + k\left(\sum_{i=1'}^{(2k+1)'} x_i\right) + k^2 x_{2k+2} \leq 2k^2 + k.$$

Moreover, it would be interesting to establish König-type theorems for other classes of graphs and *general n-colorability*. Finally, a promising direction of

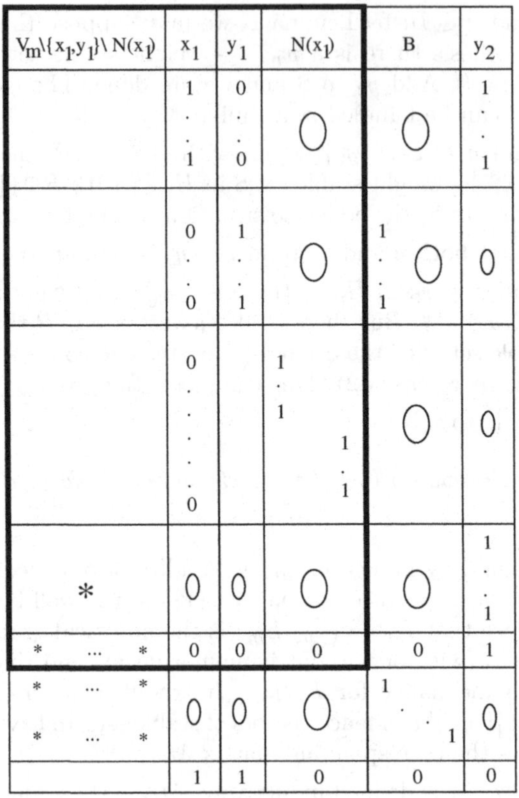

$V_m\setminus\{x_1,y_1\}\setminus N(x_1)$	x_1	y_1	$N(x_1)$	B	y_2
	1 ⋮ 1	0 ⋮ 0	O	O	1 ⋮ 1
	0 ⋮ 0	1 ⋮ 1	O	1 ⋮ 1 O	0
	0 ⋮ 0	1 ⋮ 1 1 ⋮ 1		O	0
*	0	0	O	O	1 ⋮ 1
* ... *	0	0	0	0	1
* ... * * ... *	0	0	O	1 ⋮ 1	0
	1	1	0	0	0

Fig. 5. Matrix M_{m+1}

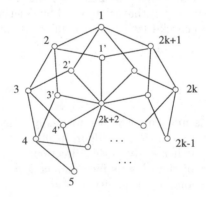

Fig. 6. Another infinite class of 4-critical graphs

research could be the generalization of Hajós' construction to hypergraphs (for which Jack Edmonds' work on matroid coloring [2] could be one of the starting points) and beyond, the study of hypergraph coloring in relation to polyhedral combinatorics.

References

1. G.Hajós. Über eine Konstruktion nicht n-färbbarer Graphen. Wiss.Z.Martin-Luther-Univ.Halle-Wittenberg, Math.-Naturw. Reihe 10, 116–117, 1961.
2. J.Edmonds. Minimum Partition of a Matroid into Independent Subsets. Journal of Research of the National Bureau of Standards 69B, 67–72, 1965.
3. V.Chvàtal. On certain polytopes associated with graphs. J.Comb.Theory B 18, 138–154, 1975.
4. T.R.Jensen B.Toft. Graph Coloring Problems. Wiley, New York, 1995.
5. R.Euler. Coloring planar Toeplitz graphs and the stable set polytope. Working paper, presented at the 17th International Symposium on Mathematical Programming, Atlanta, 2000, and the 6th International Conference on Graph Theory, Marseille, 2000 (submitted).

Algorithmic Characterization of Bipartite b-Matching and Matroid Intersection

Robert T. Firla[1*], Bianca Spille[2], and Robert Weismantel[1]

[1] Institute for Mathematical Optimization, University of Magdeburg,
Universitätsplatz 2, D-39106 Magdeburg, Germany,
{firla,weismantel}@imo.math.uni-magdeburg.de
[2] EPFL-DMA, CH-1015 Lausanne, Switzerland, bianca.spille@epfl.ch

Abstract. An algorithmic characterization of a particular combinatorial optimization problem means that there is an algorithm that works exact if and only if applied to the combinatorial optimization problem under investigation. According to Jack Edmonds, the Greedy algorithm leads to an algorithmic characterization of matroids. We deal here with the algorithmic characterization of the intersection of two matroids. To this end we introduce two different augmentation digraphs for the intersection of any two independence systems. Paths and cycles in these digraphs correspond to candidates for improving feasible solutions. The first digraph gives rise to an algorithmic characterization of bipartite b-matching. The second digraph leads to a polynomial-time augmentation algorithm for the (weighted) matroid intersection problem and to a conjecture about an algorithmic characterization of matroid intersection.

1 Introduction

This paper deals with algorithmic characterizations of some combinatorial optimization problems. It is motivated by the pioneering work of Jack Edmonds on matroids, see [5,6,7,8]. He showed that an independence system is a matroid if and only if the Greedy algorithm applied to the independence system yields an optimal solution for any weight function. This result can be understood as an algorithmic characterization of matroids with respect to independence systems. Moreover, Edmonds investigated the matroid intersection problems, characterized the corresponding polytopes and gave polynomial-time algorithms to solve the problems. We focus on an algorithmic characterization of the intersection of two matroids.

Let S be a finite set and let $\mathcal{F}$ consist of families of subsets of S. Any element F of $\mathcal{F}$ consists of a collection of subsets of S. For any $F \in \mathcal{F}$, we are interested in the family of maximization problems

$$\max c(J) : J \in F, \qquad c \in \mathbb{R}^S. \tag{1}$$

* Supported by a "Gerhard-Hess-Forschungsförderpreis" (WE 1462/2-2) of the German Science Foundation (DFG) awarded to R. Weismantel.

M. Jünger et al. (Eds.): Combinatorial Optimization (Edmonds Festschrift), LNCS 2570, pp. 48–63, 2003.

Let $D_F(J)$ be an augmentation digraph that is defined for any $F \in \mathcal{F}$, any $J \in F$ and any $c \in \mathbb{R}^S$. Its node set is $S \cup \{r, s\}$ and a node $a \in J$ has weigh c_a, a node $b \notin J$ has weight $-c_b$ and r, s have weight 0. The definition of arcs is specified in the examples. Negative (r, s)-dipaths and dicycles in this digraph correspond to candidates to augment J. We denote an (r, s)-dipath (r, P, s) by P, i.e., P consists of the inner nodes of such a sequence.

Definition 1. *An (r, s)-dipath or dicycle T in $D_F(J)$ is* feasible *for J if*

$$J \triangle T \in F.$$

A corresponding augmentation algorithm $\mathcal{A}$ can be defined as follows.

Augmentation Algorithm $\mathcal{A}$

Input: $F \in \mathcal{F}$, $J \in F$, $c \in \mathbb{R}^S$

Construct $D_F(J)$.

Find a negative feasible dicycle or (r, s)-dipath T in $D_F(J)$.

Set $J^* := J \triangle T$.

For the examples to be discussed in this paper, we may always resort to a polynomial-time algorithm to find a negative feasible dicycle or (r, s)-dipath in $D_F(J)$ when $D_F(J)$ contains such an object. Then, by the polynomial-time equivalence of augmentation and optimization for 0/1-programs [16,10], the augmentation algorithm $\mathcal{A}$ applied to F and c that uses this polynomial-time algorithm as a subalgorithm leads to a generic polynomial-time algorithm to find a locally maximal solution of (1).

In general, one may expect that $J \in F$ is not globally maximal although the corresponding augmentation digraph $D_F(J)$ does not contain a negative dicycle or (r, s)-dipath. This raises the question to characterize those $F \in \mathcal{F}$ for which the corresponding augmentation digraph D_F is *exact*.

Definition 2. *D_F is* exact *if for any $J \in F$ and any $c \in \mathbb{R}^S$, J is maximal if and only if there does not exist a negative (r, s)-dipath or dicycle in $D_F(J)$.*

Let $\mathcal{F}^*$ be a subset of $\mathcal{F}$. $\{D_F : F \in \mathcal{F}\}$ is an algorithmic characterization *of $\mathcal{F}^*$ with respect to $\mathcal{F}$ if for any $F \in \mathcal{F}$, D_F is exact if and only if $F \in \mathcal{F}^*$.*

If D_F is exact and there is a polynomial-time algorithm available to detect the negative feasible dicycles and dipaths in $D_F(J)$, then $\mathcal{A}$ leads to a polynomial-time algorithm to solve the maximization problem (1).

We mentioned that the Greedy algorithm is an algorithmic characterization of matroids with respect to independence systems. Regarding Definition 2 this can be interpreted as follows: Let $\mathcal{F}$ consist of all independence systems on S and let $\mathcal{F}^*$ consist of all matroids on S.

Definition 3. *Let $\mathcal{I}$ be an independence system on S. $\mathcal{I}$ is a* matroid *on S if for every $A \subseteq S$ every maximal independent subset of A (a basis of A) has the same cardinality.*

For an independence system $\mathcal{I}$ on S, $J \in \mathcal{I}$ and $c \in \mathbb{R}^S$, let $D_{\mathcal{I}}(J)$ be as in Fig. 1. If there exists a negative (r, s)-dipath or dicycle in $D_{\mathcal{I}}(J)$ then there also exists one of the following form:

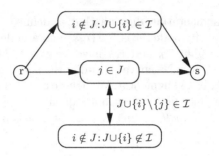

Fig. 1. The augmentation digraph $D_{\mathcal{I}}(J)$

- a negative (r, s)-dipath (b) with $b \notin J$ and $J \cup \{b\} \in \mathcal{I}$,
- a negative (r, s)-dipath (a) with $a \in J$, or
- a negative dicycle (a, b) with $a \in J, b \notin J$ and $J \cup \{b\} \setminus \{a\} \in \mathcal{I}$.

Then for a matroid $\mathcal{I}$ on S, $D_{\mathcal{I}}$ is exact. For an independence system $\mathcal{I}$ on S that is not a matroid there exists $A \subseteq S$ and a basis J of A that has not maximum cardinality. For $c = \chi^A$ (the characteristic vector of A), J is non-maximal but $D_{\mathcal{I}}(J)$ does not contain a negative (r, s)-dipath or dicycle. Hence, $D_{\mathcal{I}}$ is not exact. Therefore, $\{D_{\mathcal{I}} : \mathcal{I} \in \mathcal{F}\}$ is an algorithmic characterization of matroids ($\mathcal{F}^*$) with respect to independence systems ($\mathcal{F}$).

In this paper we deal with pairs of independence systems on S and their intersection. We define

$$\mathcal{F} := \{(\mathcal{I}_1, \mathcal{I}_2) : \mathcal{I}_1, \mathcal{I}_2 \text{ independence systems on } S\}.$$

We introduce an augmentation digraph $D(J, \mathcal{I}_1, \mathcal{I}_2) = D_{(\mathcal{I}_1, \mathcal{I}_2)}(J)$ for any two independence systems $\mathcal{I}_1$ and $\mathcal{I}_2$ on S and any common independent set J, see Fig. 2. Let $\preceq^{\mathcal{I}_1}$ and $\preceq^{\mathcal{I}_2}$ be relations on S with respect to $\mathcal{I}_1$ and $\mathcal{I}_2$, respectively. The augmentation digraph $D(J, \mathcal{I}_1, \mathcal{I}_2)$ has node set $S \cup \{r, s\}$ and arcs

$$
\begin{aligned}
&(r, i) \text{ for } i \notin J && \text{with } J \cup \{i\} \in \mathcal{I}_1; \\
&(j, i) \text{ for } j \in J, i \notin J \text{ with } J \cup \{i\} \notin \mathcal{I}_1 \text{ and } i \preceq^{\mathcal{I}_1} j; \\
&(j, s) \text{ for } j \in J; \\
&(r, j) \text{ for } j \in J; \\
&(i, j) \text{ for } j \in J, i \notin J \text{ with } J \cup \{i\} \notin \mathcal{I}_2, \text{ and } i \preceq^{\mathcal{I}_2} j; \\
&(i, s) \text{ for } i \notin J && \text{with } J \cup \{i\} \in \mathcal{I}_2.
\end{aligned}
$$

We call the first three types of arcs $\mathcal{I}_1$-*arcs* (illustrated by solid arcs) and the last three $\mathcal{I}_2$-*arcs* (dashed arcs). The arcs of any (r, s)-dipath or dicycle in this digraph alternately fulfill conditions with respect to $\mathcal{I}_1$ and $\mathcal{I}_2$.

The remaining part of this paper is divided into two sections. Every section deals with a separate family of augmentation digraphs, arising by different conditions for the relations $\preceq^{\mathcal{I}_1}$ and $\preceq^{\mathcal{I}_2}$. In Sect. 2, the relations $\preceq^{\mathcal{I}_1}$ and $\preceq^{\mathcal{I}_2}$ are quite restrictive and ensure that all (r, s)-dipaths and dicycles in $D(J, \mathcal{I}_1, \mathcal{I}_2)$ are feasible for J. The set of $\mathfrak{b}$-matchings of a bipartite graph can be represented

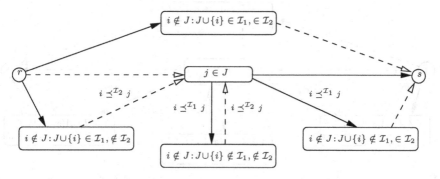

Fig. 2. The augmentation digraph $D(J, \mathcal{I}_1, \mathcal{I}_2)$

by a pair of independence systems for which the corresponding augmentation digraph is exact. Conversely, the augmentation digraph is only exact for this pair of independence systems. Hence, we obtain an algorithmic characterization of bipartite b-matching. The augmentation digraph of Sect. 3 has as subdigraph the augmentation digraph of Sect. 2. The relations $\preceq^{\mathcal{I}_1}$ and $\preceq^{\mathcal{I}_2}$ are weaker and not all dicycles or dipaths are feasible. In general, it is difficult to characterize the feasible dicycles and dipaths and hence to define the augmentation algorithm $\mathcal{A}$. In the case where both independence systems are matroids, we are able to give a polynomial-time algorithm for finding negative feasible dipaths and dicycles, we prove that the augmentation digraph is exact, and thereby we obtain a polynomial-time algorithm to solve the matroid intersection problem. Moreover, we conjecture that this augmentation digraph supplies an algorithmic characterization of matroid intersection. More precisely, we conjecture that if for two independence systems $\mathcal{I}_1$ and $\mathcal{I}_2$ on S the augmentation digraph $D_{(\mathcal{I}_1, \mathcal{I}_2)}$ is exact, then the intersection of $\mathcal{I}_1$ and $\mathcal{I}_2$ can also be represented as the intersection of two matroids defined on the same ground set S. We present several partial results that support this conjecture.

2   Algorithmic Characterization of Bipartite b-Matching

Let A and C be nonnegative integral matrices, and b, d nonnegative integral vectors such that $\mathcal{I}_1 := \{J \subseteq S : A\chi^J \leq b\}$ and $\mathcal{I}_2 := \{J \subseteq S : C\chi^J \leq d\}$. The definition of $\preceq^{\mathcal{I}_1}$ and $\preceq^{\mathcal{I}_2}$ depends on the special choice of A and C:

$$i \preceq^{\mathcal{I}_1} j : \qquad Ae^i \leq Ae^j$$
$$i \preceq^{\mathcal{I}_2} j : \qquad Ce^i \leq Ce^j$$

To be more precise, we denote the augmentation digraph that corresponds to the common independent set J by $D(J, A, C)$ or $D_{A,C}(J)$, see Fig. 3 for an illustration. The following lemma assures that all dipaths and dicycles are feasible.

Lemma 1. *Every (r, s)-dipath or dicycle in $D(J, A, C)$ is feasible for J.*

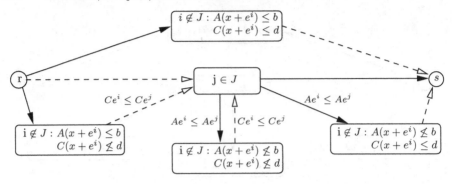

Fig. 3. The augmentation digraph $D(J, A, C)$ with $x = \chi^J$

Proof. Let T be an (r, s)-dipath or dicycle of the form $T = (i_0, j_1, i_1, \ldots, j_k, i_k)$, where i_0 and i_k are optional, depending on the structure of the dipath or dicycle. Let

$$x := \chi^J \quad \text{and} \quad t := \sum_{i \in T \setminus J} e^i - \sum_{j \in T \cap J} e^j.$$

Then $x + t = \chi^{J \triangle T}$ and we have

$$A(x + t) = \underbrace{A(x + e^{i_0})}_{\leq b} + \sum_{l=1}^{k-1} \underbrace{(Ae^{i_l} - Ae^{j_l})}_{\leq 0} + \underbrace{(Ae^{i_k} - Ae^{j_k})}_{\leq 0} \leq b.$$

Analogously, $C(x + t) = (Ce^{i_0} - Ce^{j_1}) + \sum_{l=2}^{k}(Ce^{i_{l-1}} - Ce^{j_l}) + C(x + e^{i_k}) \leq d$ and therefore, $J \triangle T$ is a common independent set of $\mathcal{I}_1$ and $\mathcal{I}_2$. $\square$

We illustrate the augmentation digraph on an example.

Example 1. Consider the intersection of two 0/1-programs,

$$\max \begin{pmatrix} 2 & 1 & 9 & 8 & 4 & 5 & 4 & 6 \end{pmatrix} x$$

$$\begin{pmatrix} 1 & 1 & & & & & & \\ & & 1 & 1 & 1 & 1 & & \\ & & & & & & 1 & 1 \end{pmatrix} x \leq \begin{pmatrix} 1 \\ 1 \\ 2 \end{pmatrix}$$

$$\begin{pmatrix} 3 & 1 & 4 & 2 & 5 & 4 & 4 & 5 \end{pmatrix} x \leq \quad 13$$

$$x \in \{0, 1\}^8.$$

The first 3×8-matrix represents the system $Ax \leq b$ and thereby the independence system $\mathcal{I}_1 = \{J \subseteq \{1, 2, \ldots, 8\} : A\chi^J \leq b\}$. The system $Cx \leq d$ corresponds in our example to the second 1×8-matrix and describes the independence system $\mathcal{I}_2 = \{J \subseteq \{1, 2, \ldots, 8\} : C\chi^J \leq d\}$. We start with the feasible solution $x := e^1 + e^3 + e^8$ that represents the common independent set $J = \{1, 3, 8\}$ and has weight 17. It may be checked that x cannot be improved by a two exchange, i.e., a vector of the form $(e^i - e^j)$ for $i \in \{2, 4, 5, 6, 7\}$ and $j \in \{1, 3, 8\}$. The

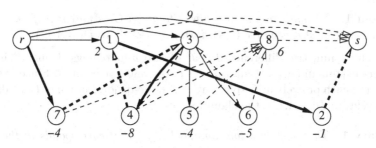

Fig. 4. $D(J, A, C)$ for $J = \{1, 3, 8\}$ and $x = e^1 + e^3 + e^8$

augmentation digraph $D(J, A, C)$ is depicted in Fig. 4. The numbers attached to the nodes correspond to the node weights. The augmentation digraph contains no negative dicycle. The (r, s)-dipath of minimal weight is $P := (7, 3, 4, 1, 2)$ with weight -2. It is represented by thick lines. Let $J' := J \bigtriangleup P = \{2, 4, 7, 8\}$ and $y := \chi^{J'} = e^2 + e^4 + e^7 + e^8$. Then y is feasible, attaining a weight of 19.

We next show that the set of b-matchings in a bipartite graph is the intersection of two independence systems $\mathcal{I}_1 = \{J \subseteq S : A\chi^J \leq b\}$ and $\mathcal{I}_2 = \{J \subseteq S : C\chi^J \leq d\}$ such that the augmentation digraph $D_{(A,C)}$ is exact. In fact, the definition of the relations $\preceq^{\mathcal{I}_1}$ and $\preceq^{\mathcal{I}_2}$ and hence of the augmentation digraph is motivated by the bipartite matching problem.

Definition 4. *Let $G = (V, E)$ be a bipartite graph with bipartition $V = V_1 \cup V_2$ and $b \in \mathbb{Z}_+^V$. $J \subseteq E$ is a b-matching in G if $|J \cap \delta(v)| \leq b_v$ for all $v \in V$. The* bipartite b-matching problem *is to find a maximal b-matching in G for any $c \in \mathbb{R}^E$.*

Let $S := E$ and for $i = 1, 2$, define

$$\mathcal{I}_i := \{J \subseteq E : |J \cap \delta(v)| \leq b_v \text{ for all } v \in V_i\}. \tag{2}$$

Then $\mathcal{I}_1$ and $\mathcal{I}_2$ are independence systems on S and their common independent sets are precisely the b-matchings of G. Let A and C be the incidence matrices of (V_1, E) and (V_2, E), respectively, $b := (b_v)_{v \in V_1}$ and $d := (b_v)_{v \in V_2}$. Then $\mathcal{I}_1 = \{J \subseteq E : A\chi^J \leq b\}$ and $\mathcal{I}_2 = \{J \subseteq E : C\chi^J \leq d\}$. We have $i \preceq^{\mathcal{I}_1} j$ if and only if the end node of i in V_1 coincides with the end node of j in V_1 and $i \preceq^{\mathcal{I}_2} j$ if and only if the end nodes of i and j in V_2 are the same.

Definition 5. *Let J be a b-matching in G. A node $v \in V$ is called J-exposed if $|J \cap \delta(v)| < b_v$. A path P (or cycle C) in G is called J-alternating if (i) the edges of P (or C) are alternately in and not in J and (ii) if the first (or last) edge of the path P is not in J, then the corresponding first (or last) node of P is J-exposed.*

Note that this definition implies that the symmetric difference of the b-matching J with a J-alternating path or cycle is again a b-matching in G.

As a slight extension of Berge's augmenting path theorem for matchings [2] we obtain

Theorem 1. *A b-matching J in a bipartite graph G is maximal if and only if there does not exist a negative (r, s)-dipath or dicycle in $D(J, A, C)$.*

Proof. The symmetric difference $J \triangle J'$ of two b-matchings J and J' in G is the edge-disjoint union of J-alternating paths and cycles in G. Since every J-alternating path or cycle in G splits into the node-disjoint union of (r, s)-dipaths and dicycles in $D(J, A, C)$, the claim follows. $\square$

Corollary 1. *The augmentation digraph $D_{(A,C)}$ that corresponds to the set of bipartite b-matchings is exact.*

We now address the converse question. If the augmentation digraph is exact for a problem, then what can we say about the problem under investigation?

For technical reasons, we require that the inequalities of $Ax \le b, Cx \le d$ define 0/1-facets of the polytope

$$\mathrm{conv}\{x \in \{0, 1\}^n : Ax \le b, \ Cx \le d\}.$$

There exist several examples that fulfill the additional assumption. For the bipartite b-matching problem, all inequalities $x(\delta(v)) \le b_v$ where $b_v < \deg(v)$ (and these are the only inequalities needed) define 0/1-facets of the bipartite b-matching polytope. Hence, for this problem, the required description is available. Moreover, any matroid intersection polytope has only 0/1-facets. Last but not least, for the stable set problem in a graph $G = (V, E)$, let P be the convex hull of all stable sets in G and let $Ax \le b, Cx \le d$ be the clique inequalities that correspond to all maximal cliques in G. Then these inequalities define 0/1-facets of $P = \mathrm{conv}\{x \in \{0, 1\}^n : Ax \le b, \ Cx \le d\}$.

Having made this further assumption, Theorem 2 implies that the exactness of the augmentation digraph for such a problem forces the problem to be a bipartite b-matching problem. This together with Corollary 1 yields an algorithmic characterization of bipartite b-matching.

Theorem 2. *Let $\mathcal{I}_1 = \{J \subseteq S : A\chi^J \le b\}$ and $\mathcal{I}_2 = \{J \subseteq S : C\chi^J \le d\}$ be two independence systems on S such that each inequality of $Ax \le b, Cx \le d$ defines a 0/1-facet of the polytope $P = \mathrm{conv}\{x \in \{0, 1\}^n : Ax \le b, \ Cx \le d\}$. If the augmentation digraph $D_{(A,C)}$ is exact then $\mathcal{I}_1 \cap \mathcal{I}_2$ is the set of b-matchings in a bipartite graph.*

Proof. Let $\mathcal{I} := \mathcal{I}_1 \cap \mathcal{I}_2 = \{J \subseteq S : \chi^J \in P\}$. Since $\mathcal{I}$ is an independence system, any inequality of $Ax \le b, Cx \le d$ has the form $x(F) \le r(F)$ with $F \subseteq S$ and $r(F)$ is the rank of F in $\mathcal{I}$. We claim that each column of A and C contains at most one nonzero entry, respectively. Assume that there are inequalities $x(F) \le r(F)$ and $x(F') \le r(F')$ of $Ax \le b$ with $F \ne F'$ and $F \cap F' \ne \emptyset$. Then

$$S := P \cap \{x : x(F) = r(F)\} \quad \text{and} \quad S' := P \cap \{x : x(F \cap F') = r(F \cap F')\}$$

define faces of P with $S \not\subseteq S'$, since S is even a facet, $F \ne F'$, and $S' \ne P$. Therefore, there exists $y \in (S \setminus S') \cap \{0, 1\}^n$. For $J := \mathrm{supp}(y) \cap F$, we have $J \in \mathcal{I}$, $J \subseteq F$, $|J| = r(F)$, and $|J \cap F'| < r(F \cap F')$. Let J^* be a maximal independent set in $F \cap F'$ and $c := \chi^{(J \cup J^*) \cap (F \cap F')}$. Then $c(J) = |J \cap F'| < r(F \cap F') = |J^*| = c(J^*)$. Hence J is not maximal. Since the augmentation digraph is exact, there exists a negative (r, s)-dipath or dicycle T in $D(J, A, C)$, i.e., we have

$$|T \cap (J \cap (F \cap F'))| < |T \cap ((J^* \setminus J) \cap (F \cap F'))|.$$

Since J is maximally independent in F it follows $J^* \setminus J \subseteq \{i \notin J : A(y + e^i) \not\leq b\}$. Thus, there exists an $\mathcal{I}_1$-arc (j, i) in T with $j \in J$, $i \notin J$ such that $c(j) = 0$ and $c(i) = 1$. This implies that $j \in J \setminus (F \cap F')$ and $i \in J^* \setminus J \subseteq F \cap F'$. We obtain $j \notin F'$ and $i \in F'$, i.e., $\chi_i^{F'} = 1$ and $\chi_j^{F'} = 0$. This proves the claim, since this means that $A_{.i} \not\leq A_{.j}$ contradicting the condition of an A-arc.

Extend the systems $Ax \leq b$ and $Cx \leq d$ by adding the upper bound inequalities $x_i \leq 1$ such that any of the new systems $A'x \leq b'$ and $C'x \leq d'$ contains exactly one 1-entry in each column. Now these systems represent the incidence matrix of a bipartite graph $G = (V_1 \cup V_2, E)$, where V_1 and V_2 correspond to the rows of A' and C', respectively. With $b := (b', d')$ the elements of $\mathcal{I}$ are exactly the b-matchings in G.   □

3 About an Algorithmic Characterization of Matroid Intersection

In this section we consider different conditions for the relations $\preceq^{\mathcal{I}_1}$ and $\preceq^{\mathcal{I}_2}$ and therefore, a different augmentation digraph than in the previous section. The conditions for the relations are motivated by the matroid intersection problem.

Definition 6. *Let $\mathcal{I}_1$ and $\mathcal{I}_2$ be matroids on S. The matroid intersection problem is to find a maximum weighted common independent set of $\mathcal{I}_1, \mathcal{I}_2$ for any $c \in \mathbb{R}^S$.*

Definition 7. *Let $\mathcal{I}$ be an independence system on S and $C, A \subseteq S$. C is a circuit of $\mathcal{I}$ if it is minimally dependent. The set of all circuits of $\mathcal{I}$ is the circuit system of $\mathcal{I}$. The rank $r(A)$ of A is the size of a maximal basis of A, whereas the lower rank $r_l(A)$ of A is the size of a minimal basis of A. The rank quotient of $\mathcal{I}$ is $\min\{r_l(A)/r(A) : A \subseteq S, r(A) > 0\}$. We denote by $\mathrm{span}(A)$ the maximal superset of A having the same rank as A.*

The rank quotient of a matroid is 1. In general, the rank quotient of an independence system is not easy to determine. It is known that if an independence system $\mathcal{I}$ on S is the intersection of m matroids on S, then the rank quotient of $\mathcal{I}$ is at least $1/m$.

The bipartite b-matching problem is a special case of the matroid intersection problem since the independence systems defined as in (2) are matroids.

We want to define the relations $\preceq^{\mathcal{I}_1}$ and $\preceq^{\mathcal{I}_2}$ in such a way that the augmentation digraph $D_{(\mathcal{I}_1, \mathcal{I}_2)}$ is exact for two matroids $\mathcal{I}_1, \mathcal{I}_2$ on S. Hence, we have to weaken the conditions for the relations of the previous section, i.e., if $i \preceq^{\mathcal{I}_1} j$ according to the relation introduced in Sect. 2 then $i \preceq^{\mathcal{I}_1} j$ according to the relation introduced in this section but now there exists more pairs of elements of S that are in relation with respect to $\preceq^{\mathcal{I}_1}$, same for the relation $\preceq^{\mathcal{I}_2}$.

Let $\mathcal{I}_1, \mathcal{I}_2$ be two independence systems on S and J a common independent set. The relations $\preceq^{\mathcal{I}_1}$ and $\preceq^{\mathcal{I}_2}$ of this section are defined as follows:

$$b \preceq^{\mathcal{I}_1} a : \qquad J \setminus \{a\} \cup \{b\} \in \mathcal{I}_1$$
$$b \preceq^{\mathcal{I}_2} a : \qquad J \setminus \{a\} \cup \{b\} \in \mathcal{I}_2$$

The corresponding augmentation digraph $D(J, \mathcal{I}_1, \mathcal{I}_2)$ is illustrated in Fig. 5.

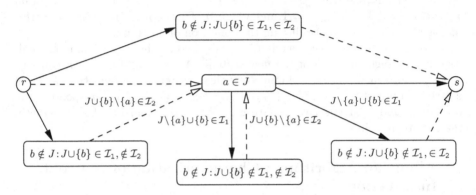

Fig. 5. The augmentation digraph $D(J, \mathcal{I}_1, \mathcal{I}_2)$

For $\mathcal{I}_1 = \mathcal{I}_2 =: \mathcal{I}$ and $J \in \mathcal{I}$, the augmentation digraph $D(J, \mathcal{I}_1, \mathcal{I}_2)$ coincides with $D_{\mathcal{I}}(J)$ defined in Sect. 1.

We remark that the difference to the augmentation digraph used in the algorithm for the cardinality matroid intersection problem or in Frank's weight-splitting algorithm for the matroid intersection problem are the additional arcs (r, a) and (a, s), see e.g. [3].

In the case of bipartite $\mathfrak{b}$-matching, let $\mathcal{I}_1$ and $\mathcal{I}_2$ be defined as in (2). Then for a $\mathfrak{b}$-matching J and $a \in J, b \notin J$ with $J \cup \{b\} \notin \mathcal{I}_1$ we have $J \cup \{b\} \setminus \{a\} \in \mathcal{I}_1$ if and only if a and b have the same end nodes in V_1, same for $\mathcal{I}_2$. Hence, the relations and the augmentation digraphs of the both sections coincide.

The following lemma implies that for all pairs of matroids $\mathcal{I}_1, \mathcal{I}_2$ on S, we can use a shortest path algorithm to find feasible negative dicycles or dipaths in $D(J, \mathcal{I}_1, \mathcal{I}_2)$ if the digraph contains any negative dipath or dicycle.

Lemma 2. *Let $\mathcal{I}_1, \mathcal{I}_2$ be matroids on S, $c \in \mathbb{R}^S$, J a common independent set, and $D(J) := D(J, \mathcal{I}_1, \mathcal{I}_2)$. Any minimal (w.r.t. cardinality) negative dicycle in $D(J)$ is feasible for J. If there does not exist a negative dicycle in $D(J)$, then any minimal shortest (r, s)-dipath in $D(J)$ is feasible for J.*

The proof of this lemma makes use of the following proposition that can be proved by induction.

Proposition 1. *[12] Let $\mathcal{I}$ be a matroid on S, $J \in \mathcal{I}$, and $b_1, a_1, \ldots, b_n, a_n$ a sequence of distinct elements of S such that*

(i) $b_i \notin J$, $a_i \in J$ for $1 \le i \le n$;
(ii) $J \cup \{b_i\} \notin \mathcal{I}$, $J \cup \{b_i\} \setminus \{a_i\} \in \mathcal{I}$ for $1 \le i \le n$;
(iii) $J \cup \{b_i\} \setminus \{a_j\} \notin \mathcal{I}$ for $1 \le i < j \le n$.

Then $J' := J \triangle \{b_1, a_1, \dots, b_n, a_n\} \in \mathcal{I}$ and $\mathrm{span}(J') = \mathrm{span}(J)$.

We next present the proof of Lemma 2. In the special case where J is maximal among all common independent sets of cardinality $|J|$ a similar proof is presented in [14] and in [9].

Proof. (of Lemma 2) Let C be a negative dicycle in $D(J)$ of minimal cardinality. Suppose there exists an undirected cycle K in $D(J)$ that consists only of $\mathcal{I}_1$-arcs and is C-alternating, meaning that its edges are alternately in and not in C, see Fig. 6 (a). We claim that there exist $k := \frac{1}{2}|K|$ dicycles $C_1, \dots, C_k$ in $D(J)$ such that their node sets are subsets of the node set of C and the sum of their weights is a nonnegative integral multiple of the weight of C, a contradiction to the minimality of C. The construction of these cycles is illustrated in Fig. 6. Let

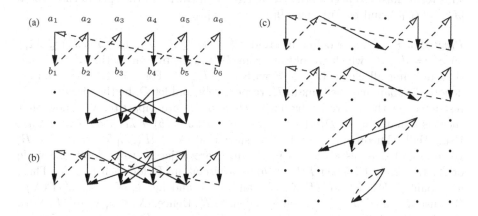

Fig. 6. (a) A cycle $C = \{a_1, b_1, \dots, a_6, b_6\}$ and a C-alternating cycle K in $D(J)$, (b) the union of C and K, (c) the cycles $C_1, \dots, C_4$ which cover each $\mathcal{I}_2$-arc 2 times

$C = (a_1, b_1, \dots, a_n, b_n)$ with $a_1, \dots, a_n \in J$, $b_1, \dots, b_n \notin J$. Let $\{d_1, \dots, d_k\}$ be the $\mathcal{I}_1$-arcs of K that are not in C. For each $d_i = (a_{i_1}, b_{i_2})$ define C_i as the unique alternating dicycle in $C \cup K$ with $C_i \cap \{d_1, \dots, d_k\} = \{d_i\}$. Then $|C_i| < |C|$ and $V(C_i) \subseteq V(C)$ for $1 \le i \le k$. We next show that every $\mathcal{I}_2$-arc of C (and hence every node of C) is contained in exactly $\lambda := |\{d_i : i_1 < i_2\}|$ dicycles of $C_1, \dots, C_k$. For (b_n, a_1) this is obvious since (b_n, a_1) is an arc of C_i if and only if $i_1 < i_2$. Let (b_j, a_{j+1}) be an $\mathcal{I}_2$-arc of C with $j < n$. Since K is a cycle, we have

$$|\{d_i : i_1 \le j < i_2\}| = |\{d_i : i_2 \le j < i_1\}|. \tag{3}$$

It is $(b_j, a_{j+1}) \in C_i$ if and only if $j+1 \le i_1 < i_2$, $i_1 < i_2 \le j$, or $i_2 \le j < i_1$. Since $|\{d_i : j+1 \le i_1 < i_2\}| + |\{d_i : i_1 < i_2 \le j\}| + |\{d_i : i_1 \le j < i_2\}| = \lambda$

with (3) follows that (b_j, a_{j+1}) is contained in exactly λ dicycles of $C_1, \ldots, C_k$. Hence, the sum of the weights of $C_1, \ldots, C_k$ is λ times the weight of C.

Therefore, there does not exist a cycle in $D(J)$ that consists only of $\mathcal{I}_1$-arcs and is C-alternating. Proposition 1 implies that $J \triangle C$ is independent in $\mathcal{I}_1$. Similarly, $J \triangle C$ is independent in $\mathcal{I}_2$. Consequently, C is feasible for J.

The proof for the (r, s)-dipath is similar. □

Lemma 2 implies that J is not maximal if there exists a negative (r, s)-dipath or dicycle in the augmentation digraph $D(J, \mathcal{I}_1, \mathcal{I}_2)$. The following decomposition result implies the existence of a negative dicycle or (r, s)-dipath in $D(J, \mathcal{I}_1, \mathcal{I}_2)$ for any non-maximal common independent set J.

Lemma 3. *Let $\mathcal{I}_1, \mathcal{I}_2$ be matroids on S, $c \in \mathbb{R}^S$, J a common independent set, and $D(J) := D(J, \mathcal{I}_1, \mathcal{I}_2)$. Let J, J^* be two common independent sets. Then $J \triangle J^*$ is the union of pairwise node-disjoint dicycles and (r, s)-dipaths in $D(J)$.*

This result has also been obtained by Krogdahl [13,14] in the special case where $|J^*| = |J| + 1$ and by Fujishige [9] for $|J^*| = |J|$.

Proof. For $i = 1, 2$, we define matroids $\mathcal{I}'_i := \{I \subseteq J \triangle J^* : I \cup (J \cap J^*) \subseteq \mathcal{M}_i\}$ on $S' := J \triangle J^*$ which we obtain from $\mathcal{I}_1$ and $\mathcal{I}_2$ by contracting $J \cap J^*$ and then deleting all elements $e \in S$ with $e \notin J \cup J^*$. For $i = 1, 2$, let span'_i, r'_i denote the span and the rank in $\mathcal{I}'_i$, respectively, and let D_i be the digraph $D(J)$ restricted on the $\mathcal{I}_i$-arcs. Obviously, D_1 and D_2 are bipartite. We show that there is a matching in D_1 of $B := \{b \in J^* \setminus J : J \cup \{b\} \notin \mathcal{I}_1\}$ into $A := J \setminus J^*$. Using Hall's Theorem [11], we have to show that $|X| \leq |\Gamma(X)|$ for every $X \subseteq B$, where $\Gamma(X)$ denotes all nodes in A which are adjacent to at least one node of X. Let $X \subseteq B$. Then $\Gamma(b) \cup \{b\}$ is a circuit in $\mathcal{I}'_1$ for each $b \in X$. Thus, $b \in \mathrm{span}'_1(\Gamma(b)) \subseteq \mathrm{span}'_1(\Gamma(X))$ for all $b \in X$ and hence, $X \subseteq \mathrm{span}'_1(\Gamma(X))$. Because $X \subseteq B \subseteq J^* \setminus J$, X is independent in $\mathcal{I}'_1$. Hence, $|X| \leq r'_1(\mathrm{span}'_1(\Gamma(X)))$. On the other hand, $r'_1(\mathrm{span}'_1(\Gamma(X))) = |\Gamma(X)|$ since $\Gamma(X)$ is independent in $\mathcal{I}'_1$ (because $\Gamma(X) \subseteq J \setminus J^*$). This results in $|X| \leq |\Gamma(X)|$. Hence, there is a matching in D_1 of B into A. We expand the matching to a matching on D_1 that covers exactly $J \triangle J^*$. Similarly, there is a matching in D_2 that covers $J \triangle J^*$. The union of these matchings on D_1 and D_2 leads to the union of node-disjoint (r, s)-dipaths and dicycles in $D(J)$ whose node set is $J \triangle J^*$, see Fig. 8. □

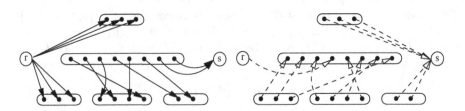

Fig. 7. A perfect matching on D_1 and D_2 restricted to the node set $J \triangle J^*$

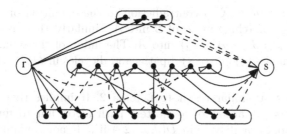

Fig. 8. $J \triangle J^*$ as the union of node-disjoint dicycles and (r, s)-dipaths in $D(J)$

Consequently, we obtain the following two equivalent statements.

Theorem 3. *Let $\mathcal{I}_1, \mathcal{I}_2$ be matroids on S and $c \in \mathbb{R}^S$. A common independent set J is maximal if and only if there does not exist a negative (r, s)-dipath or dicycle in $D(J, \mathcal{I}_1, \mathcal{I}_2)$.*

Theorem 4. *$D_{(\mathcal{I}_1, \mathcal{I}_2)}$ is exact for two matroids $\mathcal{I}_1, \mathcal{I}_2$ on S.*

Hence, we can solve the matroid intersection problem by a polynomial-time algorithm that is based on an augmentation strategy. The dipaths and dicycles T_J, that are used in each augmentation step to augment the current common independent set J, are irreducible, i.e., they are not decomposable in the sense that there does not exist T_1, T_2 such that $T_J = T_1 \cup T_2$ and $J \triangle T_1, J \triangle T_2$ are common independent sets. Fujishige [9] as well as Brezovec, Cornuejols, and Glover [1] and Camerini and Hamacher [4] presented a different digraph than $D(J, \mathcal{I}_1, \mathcal{I}_2)$ for any common independent set J to solve the augmentation problem for the matroid intersection problem. Their construction of the digraph corresponding to J depends on whether J is optimal among all independent sets of cardinality $|J|$. The dipaths and dicycles they augment with are in general not irreducible.

In the following we derive some properties of independence systems $\mathcal{I}_1, \mathcal{I}_2$ on S for which the augmentation digraph is exact. First we obtain a decomposition result similar to Lemma 3.

Theorem 5. *Let $\mathcal{I}_1, \mathcal{I}_2$ be two independence systems on S such that $D_{(\mathcal{I}_1, \mathcal{I}_2)}$ is exact. Let $J, J^* \in \mathcal{I}_1 \cap \mathcal{I}_2$ and $D(J) := D(J, \mathcal{I}_1, \mathcal{I}_2)$. Then $J \triangle J^*$ is the union of pairwise node-disjoint dicycles and (r, s)-dipaths in $D(J)$.*

Proof. For $i = 1, 2$, let D_i be the digraph which we obtain from $D(J)$ by restricting on the $\mathcal{I}_i$-arcs. Obviously, D_1 and D_2 are bipartite digraphs. We show that there is a matching in D_1 of $B := \{b \in J^* \setminus J : J \cup \{b\} \notin \mathcal{I}_1\}$ into $A := J \setminus J^*$. Using Hall's Theorem [11], we have to show that $|X| \leq |\Gamma(X)|$ for every $X \subseteq B$, where $\Gamma(X)$ denotes all nodes in A which are adjacent to at least one node of X. Let $X \subseteq B$. Suppose $|X| > |\Gamma(X)|$. Let $c := \chi^{X \cup \Gamma(X)} + \chi^{J \cap J^*}$. Then $c(J) = |\Gamma(X)| + |J \cap J^*| < |X| + |J \cap J^*| = c(J^*)$, i.e., J is not maximal. Since $D_{(\mathcal{I}_1, \mathcal{I}_2)}$ is exact there exists a negative (r, s)-dipath or dicycle Q in $D(J, \mathcal{I}_1, \mathcal{I}_2)$. Q contains for any $b \in Q \cap B$ an ingoing arc (a, b) with $a \in Q \cap J$. Since Q is negative, there exists an arc (a', b') of Q such that $c_{a'} = 0$ and $c_{b'} = 1$, i.e.,

$a' \in A \setminus \Gamma(X)$ and $b' \in X$, a contradiction to the definition of $\Gamma(X)$. Consequently, there is a matching in D_1 of B into A. Similarly, there is a matching in D_2 of $\{b \in J^* \setminus J : J \cup \{b\} \notin \mathcal{I}_2\}$ into A. The union of these matchings leads to the union of node-disjoint (r, s)-dipaths and dicycles in $D(J)$ whose node set is $J \triangle J^*$.   □

Next we show a similar result to Lemma 2. It implies that if $D_{(\mathcal{I}_1, \mathcal{I}_2)}$ is exact, we can use a variant of a shortest path algorithm to find a negative feasible (r, s)-dipath or dicycle in $D(J, \mathcal{I}_1, \mathcal{I}_2)$ if J is non-maximal. Hence, for independence systems $\mathcal{I}_1$ and $\mathcal{I}_2$ on S for which the corresponding augmentation digraph is exact we can solve the maximization problem (1) with $F = \mathcal{I}_1 \cap \mathcal{I}_2$ in polynomial-time for any $c \in \mathbb{R}^S$.

Theorem 6. *Let $\mathcal{I}_1, \mathcal{I}_2$ be two independence systems on S such that $D_{(\mathcal{I}_1, \mathcal{I}_2)}$ is exact. Let $c \in \mathbb{R}^S$, $J \in \mathcal{I}_1 \cap \mathcal{I}_2$, and $D(J) := D(J, \mathcal{I}_1, \mathcal{I}_2)$. Any minimal (w.r.t. cardinality) negative dicycle in $D(J)$ is feasible for J. If there does not exist a negative dicycle in $D(J)$, then any negative shortest (r, s)-dipath of minimal cardinality in $D(J)$ is feasible for J.*

Proof. Let $M := 2|S| \cdot \max\{1, \max\{|c_i| : i \in S\}\}$.

Let C be a minimal negative dicycle in $D(J)$. For $i \in S$,

$$c_i' := \begin{cases} -M^2 & : & i \notin J \cup C \\ M^2 & : & i \in J \setminus C \\ c_i + M & : & i \in C \end{cases}$$

The weight of C w.r.t. c' is equal to its weight w.r.t. c and hence negative. Since $D_{(\mathcal{I}_1, \mathcal{I}_2)}$ is exact, there is $J^* \in \mathcal{I}_1 \cap \mathcal{I}_2$ such that $c'(J^*) > c'(J)$. The definition of c' implies that $J \setminus C \subseteq J^* \subseteq J \cup C$, i.e., $J \triangle J^* \subseteq C$. By Theorem 5, $J \triangle J^*$ is the union of pairwise node-disjoint dicycles and (r, s)-dipaths in $D(J)$. Since no (r, s)-dipath is negative w.r.t. c' the above union contains a dicycle C' that is negative w.r.t. c'. Hence, C' is negative w.r.t. c and its node set is a subset of C. By the minimality of C, we obtain that the node sets of C and C' coincide. Consequently, $J \triangle J^* = C$, i.e., $J \triangle C = J^* \in \mathcal{I}_1 \cap \mathcal{I}_2$.

The other cases can be proved similarly.   □

Definition 8. *For sets $S_1, \ldots, S_k$ such that $S_i \not\subseteq S_j$ for $i \neq j$ we denote by $<S_1, \ldots, S_k>$ the independence system on $S_1 \cup \ldots \cup S_k$ with bases $S_1, \ldots, S_k$.*

Let $\mathcal{I}_1, \mathcal{I}_2$ be independence systems on S such that $D_{(\mathcal{I}_1, \mathcal{I}_2)}$ is exact. Then $\mathcal{I}_1$ and $\mathcal{I}_2$ need not be matroids. To see this, consider, for instance,

$$\mathcal{I}_1 = <\{1, 2\}, \{3\}, \{4\}> \quad \text{and} \quad \mathcal{I}_2 = <\{1\}, \{2\}, \{3, 4\}>.$$

In general, $D_{(\mathcal{I}_1, \mathcal{I}_2)}$ is not exact, even if $\mathcal{I}_1 \cap \mathcal{I}_2$ is a matroid. As an example, let $S = \{1, 2, 3\}$, $\mathcal{I}_1 = <\{1, 2\}, \{3\}>$, $\mathcal{I}_2 = <\{1\}, \{2, 3\}>$, and $c = 1$ be the all-ones-vector. Then $\mathcal{I}_1 \cap \mathcal{I}_2 = <\{1\}, \{2\}, \{3\}>$ is a matroid on S. $\{2\}$ is a maximal common independent set but $D(\{2\}, \mathcal{I}_1, \mathcal{I}_2)$ contains the negative (r, s)-dipath $(1, 2, 3)$. Suppose that $D_{(\mathcal{I}_1, \mathcal{I}_2)}$ is exact. Then Theorem 6 implies $\{1, 3\} \in \mathcal{I}_1 \cap \mathcal{I}_2$, a contradiction. Hence, $D_{(\mathcal{I}_1, \mathcal{I}_2)}$ is not exact.

Consequently, the fact that the intersection of two independence systems $\mathcal{I}_1$ and $\mathcal{I}_2$ on S can also be represented as the intersection of two matroids on S (or is even a matroid) does not imply the exactness of $D_{(\mathcal{I}_1,\mathcal{I}_2)}$. On the other hand, the following theorem can be interpreted as an indicator that the exactness of $D_{(\mathcal{I}_1,\mathcal{I}_2)}$ for two independence systems $\mathcal{I}_1$ and $\mathcal{I}_2$ on S forces the intersection of $\mathcal{I}_1$ and $\mathcal{I}_2$ to be an intersection of two matroids. This interpretation arises from the fact that the rank quotient of an independence system is at least $1/m$ if the independence system is the intersection of m matroids.

Theorem 7. *If $D_{(\mathcal{I}_1,\mathcal{I}_2)}$ is exact for $\mathcal{I}_1, \mathcal{I}_2$, then the rank quotient of $\mathcal{I}_1 \cap \mathcal{I}_2$ is at least $1/2$.*

Proof. Let $A \subseteq S$ and $J, J^* \in \mathcal{I}$ with $J, J^* \subseteq A$. Suppose $|J^*| > 2 \cdot |J|$. It follows $|J^* \setminus J| \geq 2 \cdot |J \setminus J^*| + 1$. Let $m := |J \setminus J^*|$. We consider subsets of $J^* \setminus J$ of cardinality $(m+1)$. Let J_1 be one of them and $J^1 := (J \cap J^*) \cup J_1$. Then $J^1 \subseteq J^* \in \mathcal{I}$ and $|J^1| = |J| + 1$. Since $D_{(\mathcal{I}_1,\mathcal{I}_2)}$ is exact, there is $i_1 \in J_1$ such that $J \cup \{i_1\} \in \mathcal{I}_1$. Let J_2 be a subset of $J^* \setminus J$ of cardinality $(m+1)$ with $i_1 \notin J_2$. Analogously follows the existence of $i_2 \in J_2$ such that $J \cup \{i_2\} \in \mathcal{I}_1$. Continuing this argumentation we obtain elements $i_1, i_2, \ldots, i_m, i_{m+1} \in J^* \setminus J$ such that $J \cup \{i_j\} \in \mathcal{I}_1$ for $1 \leq j \leq m+1$. Let $J' := (J \cap J^*) \cup \{i_1, i_2, \ldots, i_m, i_{m+1}\}$. Then $J' \subseteq J^* \in \mathcal{I}$ and $|J'| = |J| + 1$. Hence, there exists an $1 \leq j \leq m+1$ such that $J \cup \{i_j\} \in \mathcal{I}_2$ and thus, $J \cup \{i_j\} \in \mathcal{I}$. Consequently, J is not a basis of A. □

Conjecture 1. Let $\mathcal{I}_1, \mathcal{I}_2$ be independence systems on S such that $D_{(\mathcal{I}_1,\mathcal{I}_2)}$ is exact. Then the intersection of $\mathcal{I}_1$ and $\mathcal{I}_2$ can also be represented as the intersection of two matroids on S.

In general, the exactness of $D(\cdot, \mathcal{I}_1, \mathcal{I}_2)$ does not imply the existence of matroids $\mathcal{M}_1, \mathcal{M}_2$ such that $\mathcal{I}_1 \cap \mathcal{I}_2 = \mathcal{M}_1 \cap \mathcal{M}_2$ and the corresponding augmentation digraphs coincide. To see this, consider $S = \{1, 2, 3\}$, $\mathcal{I}_1 = <\{1, 2\}, \{3\}>$, and $\mathcal{I}_2 = <\{1\}, \{2\}, \{3\}>$. Then $D_{(\mathcal{I}_1,\mathcal{I}_2)}$ is exact. Suppose there exist matroids $\mathcal{M}_1, \mathcal{M}_2$ with coinciding augmentation digraphs. Then $\{1, 2\}$ and $\{3\}$ are both elements of $\mathcal{M}_1$ but $\{1, 3\}, \{2, 3\}$ are not, a contradiction to $\mathcal{M}_1$ being a matroid. This fact already implies a difficulty to prove Conjecture 1. Since the augmentation digraph of the conjectured matroids is in general different to the one of the original independence systems, it is not obvious how to define the matroids. Nevertheless, we present partial results for Conjecture 1. In particular, the validity of this conjecture has been verified in the following special cases, see [15]:

$$r(\mathcal{I}_1 \cap \mathcal{I}_2) \in \{1, 2, n-1, n\},$$
$$r(\mathcal{I}_1 \cap \mathcal{I}_2) = r_l(\mathcal{I}_1 \cap \mathcal{I}_2) = n-2, \text{ or}$$
$$n \leq 5,$$

where n is the cardinality of S. We refrain from giving the technical proofs here since they provide no insight to find a general proof. We next mention one possible attempt to prove Conjecture 1. We exhibit a construction for the supposedly existent matroids $\mathcal{M}_1$ and $\mathcal{M}_2$ on S such that $\mathcal{I}_1 \cap \mathcal{I}_2 = \mathcal{M}_1 \cap \mathcal{M}_2$. Let $\mathcal{I} := \mathcal{I}_1 \cap \mathcal{I}_2$ and let $\mathcal{C}$ be the circuit system of $\mathcal{I}$. For $i = 1, 2$, let

$$\mathcal{C}_i := \{C \in \mathcal{C} : C \notin \mathcal{I}_i\}. \tag{4}$$

We define $\mathcal{M}_i$ to be the matroid on S with the maximum number of independent sets such that any circuit in $\mathcal{C}_i$ is dependent in $\mathcal{M}_i$. Then $\mathcal{M}_1 \cap \mathcal{M}_2 \subseteq \mathcal{I}$. We conjecture that even $\mathcal{M}_1 \cap \mathcal{M}_2 = \mathcal{I} = \mathcal{I}_1 \cap \mathcal{I}_2$ holds. This is however quite difficult to prove for the following reason. It has to be shown that no element of $\mathcal{I}$ is dependent in any of the matroids $\mathcal{M}_1, \mathcal{M}_2$. We only know that $D_{(\mathcal{I}_1, \mathcal{I}_2)}$ is exact, hence, we have some information about the elements of $\mathcal{I}$ and the augmentation digraphs $D(J, \mathcal{I}_1, \mathcal{I}_2)$ for $J \in \mathcal{I}$. Consequently, proving results for $\mathcal{M}_1$ and $\mathcal{M}_2$ whose elements have in general nothing to do with any of these augmentation digraphs is a difficult task.

We consider an example. In Fig. 9 and Fig. 10 we picture two graphic matroids $\mathcal{M}_1'$ and $\mathcal{M}_2'$ on $S = \{a, b, c, d, e, f, g, h\}$. Theorem 4 guarantees the exact-

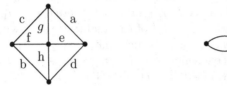

Fig. 9. Graphic matroid $\mathcal{M}_1'$ **Fig. 10.** Graphic matroid $\mathcal{M}_2'$

ness of $D_{(\mathcal{M}_1', \mathcal{M}_2')}$. Define $\mathcal{I}_1 := \mathcal{M}_1' \setminus \{abef, abgh, cdef, cdgh, efgh\}$ and $\mathcal{I}_2 := \mathcal{M}_2'$, where $s_1 s_2 \ldots s_k$ is short for $\{s_1, s_2, \ldots, s_k\}$. Then $\mathcal{I}_1 \cap \mathcal{I}_2 = \mathcal{M}_1' \cap \mathcal{M}_2'$. $\mathcal{I}_1$ is an independence system on S which is not a matroid, since aef and $abfg$ are bases of $abefg$ of different cardinality. The independent sets of $\mathcal{M}_1'$ that are not in $\mathcal{I}_1$ have no effect on the augmentation digraphs. Consequently, the family of digraphs $D_{(\mathcal{I}_1, \mathcal{I}_2)}$ is exact. The circuit system $\mathcal{C}$ of $\mathcal{I} := \mathcal{I}_1 \cap \mathcal{I}_2$ is $\mathcal{C} = \{ab, cd, ef, gh, aeg, bfh, cfg, deh\}$. Then by (4) $\mathcal{C}_1 = \{aeg, bfh, cfg, deh\}$ and $\mathcal{C}_2 = \{ab, cd, ef, gh\}$. $\mathcal{C}_2$ is the circuit system of the matroid $\mathcal{M}_2 := \mathcal{M}_2'$. $\mathcal{C}_1$ is not the circuit system of a matroid. The matroid $\mathcal{M}_1$ on S with the maximum number of independent sets such that any circuit in $\mathcal{C}_1$ is dependent in $\mathcal{M}_1$ is $\mathcal{M}_1 := \mathcal{M}_1' \cup \{abcd\}$. It is $\mathcal{M}_1 \cap \mathcal{M}_2 = \mathcal{I}_1 \cap \mathcal{I}_2$.

Finding an algorithmic characterization of matroid intersection still remains a challenging open problem.

References

1. C. Brezovec, G. Cornuejols, and F. Glover, *Two algorithms for weighted matroid intersection*, Mathematical Programming **36** (1986), 39–53.
2. C. Berge, *Two theorems in graph theory*, Proc. of the National Academy of Sciences (U.S.A.) **43** (1957), 842–844.
3. W.J. Cook, W.H. Cunningham, W.R. Pulleyblank, and A. Schrijver, *Combinatorial optimization*, Wiley-Interscience, New York, 1998.

4. P.M. Camerini and H.W. Hamacher, *Intersection of two matroids: (condensed) border graphs and ranking*, SIAM Journal on Discrete Mathematics **2** (1989), no. 1, 16–27.
5. J. Edmonds, *Matroid partition*, Math. Decision Sciences, Proceedings 5th Summer Seminary Stanford 1967, Part 1 (Lectures of Applied Mathematics 11) (1968), 335–345.
6. J. Edmonds, *Submodular functions, matroids, and certain polyhedra*, Combinatorial Structures and their Applications (R. K. Guy, H. Hanai, N. Sauer and J. Schönheim, eds.), Gordon and Brach, New York (1970), 69–87.
7. J. Edmonds, *Matroids and the greedy algorithm*, Mathematical Programming **1** (1971), 127–136.
8. J. Edmonds, *Matroid intersection*, Annals of Discrete Mathematics **4** (1979), 39–49.
9. S. Fujishige, *A primal approach to the independent assignment problem*, Journal of the Operations Research Society of Japan **20** (1977), 1–15.
10. M. Grötschel and L. Lovász, *Combinatorial optimization*, Handbook of Combinatorics (R. Graham, M. Grötschel, and L. Lovász, eds.), North-Holland, Amsterdam, 1995, pp. 1541–1598.
11. P. Hall, *On representatives of subsets*, Journal of the London Mathematical Society **10** (1935), 26–30.
12. M. Iri and N. Tomizawa, *An algorithm for finding an optimal "independent assignment"*, Journal of the Operations Research Society of Japan **19** (1976), 32–57.
13. S. Krogdahl, *A combinatorial proof of lawler's matroid intersection algorithm*, unpublished manuscript (partly published in [14]) (1975).
14. E.L. Lawler, *Combinatorial optimization: Networks and matroids*, Holt, Rinehart and Winston, New York etc., 1976.
15. B. Spille, *Primal characterizations of combinatorial optimization problems*, PhD Thesis, University Magdeburg, Germany, 2001.
16. A.S. Schulz, R. Weismantel, and G.M. Ziegler, *0/1 integer programming: optimization and augmentation are equivalent*, Algorithms-ESA95 (P. Spirakis, ed.), Lecture Notes in Computer Science 979, Springer, Berlin, 1995, pp. 473–483.

Solving Real-World ATSP Instances by Branch-and-Cut

Matteo Fischetti[1], Andrea Lodi[2] and Paolo Toth[2]

[1] DEI, University of Padova
Via Gradenigo 6/A, 35100 Padova, Italy
matteo.fischetti@unipd.it
[2] DEIS, University of Bologna
Viale Risorgimento 2, 40136 Bologna, Italy
{alodi,ptoth}@deis.unibo.it

Abstract. Recently, Fischetti, Lodi and Toth [15] surveyed exact methods for the *Asymmetric Traveling Salesman Problem* (ATSP) and computationally compared branch-and-bound and branch-and-cut codes. The results of this comparison proved that branch-and-cut is the most effective method to solve hard ATSP instances. In the present paper the branch-and-cut algorithms by Fischetti and Toth [17] and by Applegate, Bixby, Chvátal and Cook [2] are considered and tested on a set of 35 real-world instances including 16 new instances recently presented in [12].

1   Introduction

Let $G = (V, A)$ be a given complete digraph, where $V = \{1, \ldots, n\}$ is the vertex set and $A = \{(i, j) : i, j \in V\}$ the arc set, and let c_{ij} be the cost associated with arc $(i, j) \in A$ (with $c_{ii} = +\infty$ for each $i \in V$). A *Hamiltonian circuit* (*tour*) of G is a circuit visiting each vertex of V exactly once.

The *Asymmetric Traveling Salesman Problem* (ATSP) is to find a Hamiltonian circuit $G^* = (V, A^*)$ of G whose $cost \sum_{(i,j) \in A^*} c_{ij}$ is a minimum, and can be formulated as the following Integer Linear Program (ILP):

$$v(\text{ATSP}) \quad = \min \sum_{(i,j) \in A} c_{ij}\, x_{ij} \tag{1}$$

subject to

$$\sum_{i \in V} x_{ij} = 1 \qquad j \in V \tag{2}$$

$$\sum_{j \in V} x_{ij} = 1 \qquad i \in V \tag{3}$$

$$\sum_{i \in S} \sum_{j \in S} x_{ij} \leq |S| - 1 \qquad S \subset V : S \neq \emptyset \tag{4}$$

$$x_{ij} \geq 0 \qquad (i, j) \in A \tag{5}$$

$$x_{ij} \text{ integer} \qquad (i, j) \in A \tag{6}$$

where $x_{ij} = 1$ if and only if arc (i, j) is in the optimal tour. Constraints (2) and (3) impose that the in-degree and out-degree of each vertex, respectively, is equal to one, while constraints (4) are *Subtour Elimination Constraints* (SECs) and impose that no partial circuit exists.

M. Jünger et al. (Eds.): Combinatorial Optimization (Edmonds Festschrift), LNCS 2570, pp. 64–77, 2003.
© Springer-Verlag Berlin Heidelberg 2003

To simplify notation, for any $f : A \to \mathbb{R}$ and $S_1, S_2 \subseteq V$, we write $f(S_1, S_2)$ for $\sum_{i \in S_1} \sum_{j \in S_2} f_{ij}$; moreover, we write $f(i, S_2)$ or $f(S_1, i)$ whenever $S_1 = \{i\}$ or $S_2 = \{i\}$, respectively.

Fischetti, Lodi and Toth [15] recently surveyed exact methods for ATSP and computationally compared four exact codes, namely, the branch-and-bound code by Carpaneto, Dell'Amico and Toth [11] based on the *Assignment Problem* (AP) relaxation[1], the additive branch-and-bound code by Fischetti and Toth [16], the branch-and-cut code by Fischetti and Toth [17], and the branch-and-cut *Symmetric* TSP (STSP) code by Applegate, Bixby, Chvátal and Cook [1]. These codes have been extensively tested on a set of 86 ATSP instances. The results of this comparison proved that branch-and-cut is the most effective method to solve hard ATSP instances.

In the present paper the two branch-and-cut codes considered in [15] are tested on a set of 35 real-world instances including 16 new instances recently presented in [12]. The paper is organized as follows. In Section 2 the branch-and-cut method by Fischetti and Toth [17] is briefly discussed, while in Section 3 the adaptations to the STSP branch-and-cut code by Applegate, Bixby, Chvátal and Cook [1] needed to solve ATSP instances are presented. In Section 4 the codes are compared on the set of instances. Some conclusions are finally drawn in Section 5.

2   The Branch-and-Cut Algorithm by Fischetti and Toth

We next outline the polyhedral method of Fischetti and Toth [17]. Branch-and-cut methods for ATSP with side constraints have been proposed recently by Ascheuer [3], Ascheuer, Jünger and Reinelt [6], and Ascheuer, Fischetti and Grötschel [4,5], among others. The Fischetti-Toth method is based on model (1)–(6), and exploits additional classes of facet-inducing inequalities for the ATSP polytope P that proved to be of crucial importance for the solution of some real-world instances. For each class, one needs to address the associated *separation* problem (in its optimization version), defined as follows: Given a point $x^* \geq 0$ satisfying the degree equations, along with a family $\mathcal{F}$ of ATSP inequalities, find a most violated member of $\mathcal{F}$, i.e., an inequality $\alpha x \leq \alpha_0$ belonging to $\mathcal{F}$ and maximizing the degree of violation $\alpha x^* - \alpha_0$.

2.1   Separation of Symmetric Inequalities

An ATSP inequality $\alpha x \leq \alpha_0$ is called *symmetric* when $\alpha_{ij} = \alpha_{ji}$ for all $(i, j) \in A$. Symmetric inequalities can be thought of as derived from valid inequalities for the STSP, defined as the problem of finding a minimum-cost Hamiltonian cycle in a given undirected graph $G_E = (V, E)$. Indeed, let $y_e = 1$ if edge $e \in E$ belongs to the optimal STSP solution; $y_e = 0$ otherwise. Every inequality $\sum_{e \in E} \alpha_e y_e \leq \alpha_0$ for STSP can be transformed into a valid ATSP inequality by

---

[1] The AP is the relaxation obtained by dropping constraints (4).

simply replacing y_e by $x_{ij} + x_{ji}$ for all edges $e = (i, j) \in E$. This produces the symmetric inequality $\alpha x \leq \alpha_0$, where $\alpha_{ij} = \alpha_{ji} = \alpha_{(i,j)}$ for all $i, j \in V$, $i \neq j$. Conversely, every symmetric ATSP inequality $\alpha x \leq \alpha_0$ corresponds to the valid STSP inequality $\sum_{(i,j)\in E} \alpha_{ij} y_{(i,j)} \leq \alpha_0$.

The above correspondence implies that every separation algorithm for STSP can be used, as a "black box", for ATSP as well. To this end, given the ATSP (fractional) point x^* one first defines the undirected counterpart y^* of x^* by means of the transformation

$$y_e^* := x_{ij}^* + x_{ji}^* \quad \text{for all edges } e = (i, j) \in E$$

and then applies the STSP separation algorithm to y^*. On return, the detected most violated STSP inequality is transformed into its ATSP counterpart, both inequalities having the same degree of violation.

Several exact/heuristic separation algorithms for STSP have been proposed in recent years, all of which can be used for ATSP. In [17] only two such separation tools are used, namely: (i) the Padberg-Rinaldi [26] exact algorithm for SECs; and (ii) the simplest heuristic scheme for comb (actually, 2-matching) constraints, in which the components of the graph induced by the edges $e \in E$ with fractional y_e^* are considered as potential handles of the comb.

2.2   Separation of D_k^+ and D_k^- Inequalities

The following D_k^+ inequalities have been proposed by Grötschel and Padberg [19]:

$$x_{i_1 i_k} + \sum_{h=2}^{k} x_{i_h i_{h-1}} + 2 \sum_{h=2}^{k-1} x_{i_1 i_h} + \sum_{h=3}^{k-1} x(\{i_2, \dots, i_{h-1}\}, i_h) \leq k - 1 \quad (7)$$

where $(i_1, \dots, i_k)$ is any sequence of $k \in \{3, \dots, n - 1\}$ distinct vertices. D_k^+ inequalities are facet-inducing for the ATSP polytope [13], and are obtained by lifting the cycle inequality $\sum_{(i,j)\in C} x_{ij} \leq k - 1$ associated with the subtour $C := \{(i_1, i_k), (i_k, i_{k-1}), \dots, (i_2, i_1)\}$. The separation problem for the class of D_k^+ inequalities calls for a vertex sequence $(i_1, \dots, i_k)$, $3 \leq k \leq n-1$, for which the degree of violation $x_{i_1 i_k}^* + \sum_{h=2}^{k} x_{i_h i_{h-1}}^* + 2 \sum_{h=2}^{k-1} x_{i_1 i_h}^* + \sum_{h=3}^{k-1} x^*(\{i_2, \dots, i_{h-1}\}, i_h)$ $-k + 1$ is as large as possible. This is itself a combinatorial optimization problem that can be effectively solved in practice by an implicit enumeration scheme enhanced by suitable pruning conditions [17].

Also addressed in [17] are the following D_k^- inequalities:

$$x_{i_k i_1} + \sum_{h=2}^{k} x_{i_{h-1} i_h} + 2 \sum_{h=2}^{k-1} x_{i_h i_1} + \sum_{h=3}^{k-1} x(i_h, \{i_2, \dots, i_{h-1}\}) \leq k - 1 \quad (8)$$

where $(i_1, \dots, i_k)$ is any sequence of $k \in \{3, \dots, n - 1\}$ distinct nodes. D_k^- inequalities are valid [19] and facet-inducing [13] for P; they can be obtained by lifting the cycle inequality $\sum_{(i,j)\in C} x_{ij} \leq k - 1$ associated with the circuit $C := \{(i_1, i_2), \ldots, (i_{k-1}, i_k), (i_k, i_1)\}$.

D_k^- inequalities can be thought of as derived from D_k^+ inequalities by swapping the coefficient of the two arcs (i, j) and (j, i) for all $i, j \in V$, $i < j$. This is a perfectly general operation, called *transposition* in [19], that works as follows.

For every $\alpha \in \mathbb{R}^A$, let $\alpha^T \in \mathbb{R}^A$ be defined by: $\alpha_{ij}^T := \alpha_{ji}$ for all $(i, j) \in A$. Clearly, inequality $\alpha x \leq \alpha_0$ is valid (or facet-inducing) for the ATSP polytope P if and only if its transposed version, $\alpha^T x \leq \alpha_0$, is. This follows from the obvious fact that $\alpha^T x = \alpha x^T$, where $x \in P$ if and only if $x^T \in P$. Moreover, every separation procedure for $\alpha x \leq \alpha_0$ can also be used, as a black box, to deal with $\alpha^T x \leq \alpha_0$. To this end one gives the transposed point $(x^*)^T$ (instead of x^*) on input to the procedure, and then transposes the returned inequality.

The above considerations show that both the heuristic and exact separation algorithms designed for D_k^+ inequalities can be used for D_k^- inequalities as well.

2.3   Separation of Odd CAT Inequalities

The following class of inequalities has been proposed by Balas [7]. Two distinct arcs (i, j) and (u, v) are called *incompatible* if $i = u$, or $j = v$, or $i = v$ and $j = u$; *compatible* otherwise. A *Closed Alternating Trail* (CAT, for short) is a sequence $T = \{a_1, \ldots, a_t\}$ of t distinct arcs such that, for $k = 1, \ldots, t$, arc a_k is incompatible with arcs a_{k-1} and a_{k+1}, and compatible with all other arcs in T (with $a_0 := a_t$ and $a_{t+1} := a_1$). Let $\delta^+(v)$ and $\delta^-(v)$ denote the set of the arcs of G leaving and entering any vertex $v \in V$, respectively. Given a CAT T, a node v is called a *source* if $|\delta^+(v) \cap T| = 2$, whereas it is called a *sink* if $|\delta^-(v) \cap T| = 2$. Notice that a node can play both source and sink roles. Let Q be the set of the arcs $(i, j) \in A \setminus T$ such that i is a source and j is a sink node. For any CAT of odd length t, the following *odd CAT inequality*

$$\sum_{(i,j)\in T\cup Q} x_{ij} \leq \frac{|T| - 1}{2} \tag{9}$$

is valid and facet-defining (except in two pathological cases arising for $n \leq 6$) for the ATSP polytope [7].

The following heuristic separation algorithm is based on the known fact that odd CAT inequalities correspond to odd cycles on an auxiliary "incompatibility" graph [7]. Given the point x^*, we set-up an edge-weighted undirected graph $\tilde{G} = (\tilde{N}, \tilde{E})$ having a node ν_a for each arc $a \in A$ with $x_a^* > 0$, and an edge $e = (\nu_a, \nu_b)$ for each pair a, b of incompatible arcs, whose weight is defined as $w_e := 1 - (x_a^* + x_b^*)$. We assume that x^* satisfies all degree equations as well as all trivial SECs of the form $x_{ij} + x_{ji} \leq 1$; this implies $w_e \geq 0$ for all $e \in \tilde{E}$.

Let $\tilde{\delta}(v)$ contain the edges in $\tilde{E}$ incident with a given node $v \in \tilde{N}$. A *cycle* $\tilde{C}$ of $\tilde{G}$ is an edge subset of $\tilde{E}$ inducing a connected subdigraph of $\tilde{G}$, and such that $|\tilde{C} \cap \tilde{\delta}(v)|$ is even for all $v \in \tilde{N}$. Cycle $\tilde{C}$ is called (i) *odd* if $|\tilde{C}|$ is odd; (ii) *simple* if $|\tilde{C} \cap \tilde{\delta}(v)| \in \{0, 2\}$ for all $v \in \tilde{N}$; and (iii) *chordless* if the subdigraph of $\tilde{G}$ induced by the nodes covered by $\tilde{C}$ has no other edges than those in $\tilde{C}$.

By construction, every simple and chordless odd cycle $\tilde{C}$ in $\tilde{G}$ corresponds to an odd CAT T, where $a \in T$ if and only if ν_a is covered by $\tilde{C}$. In addition, the total weight of $\tilde{C}$ is $w(\tilde{C}) := \sum_{e \in \tilde{C}} w_e = \sum_{(\nu_a, \nu_b) \in \tilde{C}} (1 - x_a^* - x_b^*) = |T| - 2 \sum_{a \in T} x_a^*$ hence $(1 - w(\tilde{C}))/2$ gives a lower bound on the degree of violation of the corresponding CAT inequality, computed as $\phi(T) := (2 \sum_{a \in T \cup Q} x_a^* - |T| + 1)/2$.

The heuristic separation algorithm used in [17] computes, for each $e \in \tilde{E}$, a minimum-weight odd cycle $\tilde{C}_e$ that uses edge e. If $\tilde{C}_e$ happens to be simple and chordless, then it corresponds to an odd CAT, say T. If, in addition, the lower bound $(1 - w(\tilde{C}_e))/2$ exceeds a given threshold $\theta = -1/2$, then the corresponding inequality is hopefully violated; hence one evaluates its actual degree of violation, $\phi(T)$, and stores the inequality if $\phi(T) > 0$. In order to avoid detecting twice the same inequality, edge e is removed from $\tilde{G}$ after the computation of each $\tilde{C}_e$.

The key point of the algorithm is the computation in $\tilde{G}$ of a minimum-weight odd cycle going through a given edge. Assuming that the edge weights are all nonnegative, this problem is known to be polynomially solvable as it can be transformed into a shortest path problem; see Gerards and Schrijver [18]. To this end one constructs an auxiliary bipartite undirected graph $G_B = (N_B' \cup N_B'', E_B)$ obtained from $\tilde{G}$ as follows. For each ν in $\tilde{G}$ there are two nodes in G_B, say ν' and ν''. For each edge $e = (\nu_1, \nu_2)$ of $\tilde{G}$ there are two edges in G_b, namely edge (ν_1', ν_2'') and edge (ν_2', ν_1''), both having weight w_e. By construction, every minimum-weight odd cycle $\tilde{C}_e$ of $\tilde{G}$ going through edge $e = (\nu_1, \nu_2)$ corresponds in G_B to a shortest path from ν_1' to ν_2', plus the edge (ν_2', ν_1''). Hence, the computation of all $\tilde{C}_e$'s can be performed efficiently by computing, for each ν_1', the shortest path from ν_1' to all other nodes in N_B'.

2.4 Clique Lifting and Shrinking

Clique lifting can be described as follows, see Balas and Fischetti [9] for details. Let $P(G')$ denote the ATSP polytope associated with a given complete digraph $G' = (V', A')$. Given a valid inequality $\beta y \le \beta_0$ for $P(G')$, we define

$$\beta_{hh} := \max\{\beta_{ih} + \beta_{hj} - \beta_{ij} : i, j \in V' \setminus \{h\}, i \ne j\} \quad \text{for all } h \in V'$$

and construct an enlarged complete digraph $G = (V, A)$ obtained from G' by replacing each node $h \in V'$ by a clique S_h containing at least one node (hence, $|V| = \sum_{h \in V'} |S_h| \ge |V'|$). In other words $(S_1, \ldots, S_{|V'|})$ is a proper partition of V, in which the h-th set corresponds to the h-th node in V'.

For all $v \in V$, let $v \in S_{h(v)}$. We define a new *clique lifted* inequality for $P(G)$, say $\alpha x \le \alpha_0$, where $\alpha_0 := \beta_0 + \sum_{h \in V'} \beta_{hh}(|S_h| - 1)$ and $\alpha_{ij} := \beta_{h(i)h(j)}$ for each $(i, j) \in A$. It is shown in [9] that the new inequality is always valid for $P(G)$; in addition, if the starting inequality $\beta x \le \beta_0$ defines a facet of $P(G')$, then $\alpha x \le \alpha_0$ is guaranteed to be facet-inducing for $P(G)$.

Clique lifting is a powerful theoretical tool for extending known classes of inequalities. Also, it has important applications in the design of separation algorithms in that it allows one to simplify the separation problem through the following *shrinking* procedure [27].

Let $S \subset V$, $2 \leq |S| \leq n - 2$, be a vertex subset saturated by x^*, in the sense that $x^*(S, S) = |S| - 1$, and suppose S is shrunk into a single node, say σ, and x^* is updated accordingly. Let $G' = (V', A')$ denote the shrunken digraph, where $V' := V \setminus S \cup \{\sigma\}$, and let y^* be the shrunken counterpart of x^*. Every valid inequality $\beta y \leq \beta_0$ for $P(G')$ that is violated by y^* corresponds in G to a violated inequality, say $\alpha x \leq \alpha_0$, obtained through clique lifting by replacing back σ with the original set S. As observed by Padberg and Rinaldi [27], however, this shrinking operation can affect the possibility of detecting violated cuts on G', as it may produce a point y^* belonging to $P(G')$ even when $x^* \notin P(G)$.

The above observation shows that shrinking has to be applied with some care. There are however simple conditions on the choice of S that guarantee $y^* \notin P(G')$, provided $x^* \notin P(G)$ as in the cases of interest for separation. The simplest such condition concerns the shrinking of *1-arcs* (i.e., arcs (i, j) with $x^*_{ij} = 1$), and requires $S = \{i, j\}$ for a certain node pair i, j with $x^*_{ij} = 1$.

In [17] 1-arc shrinking is applied iteratively, so as to replace each path of 1-arcs by a single node. As a result of this pre-processing on x^*, all the nonzero variables are fractional. Notice that a similar result cannot be obtained for the symmetric TSP, where each chain of 1-edges (i.e., edges whose associated x^* variable takes value 1) can be replaced by a single 1-edge, but not by a single node.

2.5   Pricing with Degeneracy

Pricing is an important ingredient of branch-and-cut codes, in that it allows one to effectively handle LP problems involving a huge number of columns. Let

$$z := \min\{cx : Mx \equiv b, x \geq 0\} \tag{10}$$

be the LP problem to be solved. M is an $m \times |A|$ matrix whose columns are indexed by the arcs $(i, j) \in A$. The first $2n - 1$ rows of M correspond to the degree equations (2)-(3) (with the redundant constraint $x(1, V) = 1$ omitted), whereas the remaining rows, if any, correspond to some of the cuts generated through separation. Notation "$\equiv$" stands for "$=$" for the first $2n - 1$ rows of M, and "$\leq$" for the remaining rows. Let M^h_{ij} denote the entry of M indexed by row h and column (i, j).

In order to keep the size of the LP as small as possible, the following *pricing* scheme is commonly used. We determine a (small) *core* set of arcs, say $\tilde{A}$, and decide to temporarily fix $x_{ij} = 0$ for all $(i, j) \in A \setminus \tilde{A}$. We then solve the restricted LP problem

$$\tilde{z} := \min\{\tilde{c}\tilde{x} : \tilde{M}\tilde{x} \equiv b, \tilde{x} \geq 0\} \tag{11}$$

where $\tilde{c}$, $\tilde{x}$, and $\tilde{M}$ are obtained from c, x, and M, respectively, by removing all entries indexed by $A \setminus \tilde{A}$.

Assume problem (11) is feasible, and let $\tilde{x}^*$ and $\tilde{u}^*$ be the optimal primal and dual basic solutions found, respectively. Clearly, $\tilde{z} \geq z$. We are interested in easily-checkable conditions that guarantee $\tilde{z} = z$, thus proving that $\tilde{x}^*$ (with $\tilde{x}_{ij}^* := 0$ for all $(i, j) \in A \setminus \tilde{A}$) is an optimal basic solution to (10), and hence that its value $\tilde{z}$ is a valid lower bound on $v(ATSP)$. To this end we compute the LP reduced costs associated with $\tilde{u}^*$, namely

$$\bar{c}_{ij} := c_{ij} - \sum_{h=1}^m M_{ij}^h \tilde{u}_h^* \quad \text{for } (i, j) \in A$$

and check whether $\bar{c}_{ij} \geq 0$ for all $(i, j) \in A$. If this is indeed the case, then $\tilde{z} = z$ and we are done. Otherwise, the current core set $\tilde{A}$ is enlarged by adding (some of) the arcs with negative reduced cost, and the whole procedure is iterated. This iterative solution of (11), followed by the possible updating of $\tilde{A}$, is generally referred to as the *pricing loop*.

According to common computational experience, the first iterations of the pricing loop tend to add a very large number of new columns to the LP even when $\tilde{z} = z$, due to the typically high primal degeneracy of (11).

A different technique, called *AP pricing* in [17], exploits the fact that any feasible solution to (10) cannot select the arcs with negative reduced cost in an arbitrary way, as the degree equations —among other constraints— have to be fulfilled. The technique is related to the so-called *Lagrangian pricing* introduced independently by Löbel [23] as a powerful method for solving large-scale vehicle scheduling problems.

Let us consider the dual solution $\tilde{u}^*$ to (11) as a vector of Lagrangian multipliers, and the LP reduced costs $\bar{c}_{ij}$ as the corresponding Lagrangian costs. In this view, standard pricing consists of solving the following trivial relaxation of (10):

$$LB_1 := \min_{x \geq 0}[cx + \tilde{u}^*(b - Mx)] = \tilde{u}^* b + \min_{x \geq 0} \bar{c}x \tag{12}$$

where $\tilde{u}^* b = \tilde{z}$ by LP duality. Therefore one has $\tilde{z} + \min_{x \geq 0} \bar{c}x \leq z \leq \tilde{z}$, from which $\tilde{z} = z$ in case $\min_{x \geq 0} \bar{c}x = 0$, i.e., $\bar{c}_{ij} \geq 0$ for all i, j. The strengthening then consists in replacing condition $x \geq 0$ in (12) by

$$x \in F(\text{AP}) := \{x \in \{0, 1\}^A : x(i, V) = x(V, i) = 1 \text{ for all } i \in V\}$$

In this way one computes an improved lower bound on z, namely

$$LB_2 := \tilde{u}^* b + \min_{x \in F(\text{AP})} \bar{c}x = \tilde{z} + \Delta_{\text{AP}}$$

where $\Delta_{\text{AP}} := \min_{x \in F(\text{AP})} \bar{c}x$ is computed efficiently by solving the AP on the Lagrangian costs $\bar{c}_{ij}$. As before, $\tilde{z} + \Delta_{\text{AP}} \leq z \leq \tilde{z}$, hence $\Delta_{\text{AP}} = 0$ implies $\tilde{z} = z$. When $\Delta_{\text{AP}} < 0$, instead, one has to iterate the procedure, after having added to the core set $\tilde{A}$ the arcs in $A \setminus \tilde{A}$ that are selected in the optimal AP solution found.

The AP pricing has two main advantages over the standard one, namely: (1) an improved check for proving $\tilde{z} = z$; and (2) a better rule to select the arcs to be added to the core arc set. Moreover, LB_2 always gives a lower bound on z (and hence on $v(ATSP)$), which can in some cases succeed in fathoming the current branching node even when $\Delta_{AP} < 0$. Finally, the nonnegative AP reduced cost vector $\bar{\bar{c}}$ available after solving $\min_{x \in F(AP)} \bar{c}x$ can be used for fixing $x_{ij} = 0$ for all $(i, j) \in A$ such that $LB_2 + \bar{\bar{c}}_{ij}$ is at least as large as the value of the best known ATSP solution.

2.6 The Overall Algorithm

The algorithm is a lowest-first branch-and-cut procedure. At each node of the branching tree, the LP relaxation is initialized by taking all the constraints present in the last LP solved at the father node (for the root node, only the degree equations are taken). As to variables, one retrieves from a scratch file the optimal basis associated with the last LP solved at the father node, and initializes the core variable set, $\tilde{A}$, by taking all the arcs belonging to this basis (for the root node, $\tilde{A}$ contains the $2n - 1$ variables in the optimal AP basis found by solving AP on the original costs c_{ij}). In addition, $\tilde{A}$ contains all the arcs of the best known ATSP solution. Starting with the above advanced basis, one iteratively solves the current LP, applies the AP pricing (and variable fixing) procedure described in Section 2.5, and repeats if needed. Observe that the pricing/fixing procedure is applied after each LP solution.

On exit of the pricing loop (case $\Delta_{AP} = 0$), the cuts whose associated slack exceeds 0.01 are removed from the current LP (unless the number of these cuts is less than 10), and the LP basis is updated accordingly. Moreover, separation algorithms are applied to find, if any, facet-defining ATSP inequalities that cut off the current LP optimal solution, say x^*. As a heuristic rule, the violated cuts with degree of violation less than 0.1 (0.01 for SECs) are skipped, and the separation phase is interrupted as soon as $20 + \lfloor n/5 \rfloor$ violated cuts are found.

One first checks for violation the cuts generated during the processing of the current or previous nodes, all of which are stored in a global data-structure called the constraint *pool*. If some of these cuts are indeed violated by x^*, the separation phase ends. Otherwise, the Padberg-Rinaldi [26] MINCUT algorithm for SEC separation is applied, and the separation phase is interrupted if violated SECs are found. When this is not the case, one shrinks the 1-arc paths of x^* (as described in Section 2.4), and applies the separation algorithms for comb (Section 2.1), D_k^+ and D_k^- (Section 2.2), and odd CAT (Section 2.3) inequalities. In order to avoid finding equivalent inequalities, D_3^- inequalities (which are the same as D_3^+ inequalities), are never separated, and odd CAT separation is skipped when a violated comb is found (as the class of comb and odd CAT inequalities overlap). When violated cuts are found, one adds them to the current LP, and repeats.

When separation fails and x^* is integer, the current best ATSP solution is updated, and a backtracking step occurs. If x^* is fractional, instead, the current LP basis is saved in a file, and one branches on the variable x_{ij} with $0 < x_{ij}^* < 1$ that maximizes the score $\sigma(i, j) := c_{ij} \cdot \min\{x_{ij}^*, 1 - x_{ij}^*\}$. As a heuristic rule, a large priority is given to the variables with $0.4 \leq x_{ij}^* \leq 0.6$ (if any), so as to produce a significant change in both descending nodes.

As a heuristic tailing-off rule, one also branches when the current x^* is fractional and the lower bound did not increase in the last 5 (10 for the root node) LP/pricing/separation iterations.

In addition, a simple heuristic algorithm is used to hopefully update the current best optimal ATSP solution. The algorithm is based on the information associated with the current LP, and consists of a complete enumeration of the Hamiltonian circuits in the support graph of x^*, defined as $G^* := (V, \{(i, j) \in A : x_{ij}^* > 0\})$. To this end Martello's [24] implicit enumeration algorithm HC is used, with at most $100 + 10n$ backtracking steps allowed. As G^* is typically very sparse, this upper bound on the number of backtrackings is seldom attained, and HC almost always succeeds in completing the enumeration within a short computing time. The heuristic is applied whenever SEC separation fails, since in this case G^* is guaranteed to be strongly connected.

3   Using a STSP Code for ATSP Instances

It is easy to see that a code for ATSP can be invoked to solve symmetric TSP instances. In fact, the reverse also holds by means of the following two transformations:

- the *3-node* transformation proposed by Karp [22]. A complete *undirected* graph with $3n$ vertices is obtained from the original complete *directed* one by adding two copies, $n + i$ and $2n + i$, of each vertex $i \in V$, and by (i) setting to 0 the cost of the edges $(i, n + i)$ and $(n + i, 2n + i)$ for each $i \in V$, (ii) setting to c_{ij} the cost of edge $(2n + i, j)$ $\forall i, j \in V$, and (iii) setting to $+\infty$ the costs of all the remaining edges;
- the *2-node* transformation proposed by Jonker and Volgenant [20] (see also Jünger, Reinelt and Rinaldi [21]). A complete *undirected* graph with $2n$ vertices is obtained from the original complete *directed* one by adding a copy, $n+i$, of each vertex $i \in V$, and by (i) setting to 0 the cost of the edge $(i, n+i)$ for each $i \in V$, (ii) setting to $c_{ij} + M$ the cost of edge $(n + i, j)$ $\forall i, j \in V$, where M is a sufficiently large positive value, and (iii) setting to $+\infty$ the costs of all the remaining edges. The transformation value nM has to be subtracted from the STSP optimal cost.

The most effective branch-and-cut algorithm for the STSP is currently the one by Applegate, Bixby, Chvátal and Cook [2], and the corresponding code, Concorde [1], is publicly available. As already done in [15], we used this code to test the effectiveness of the approach based on the ATSP-to-STSP transformation. The code has been used with default parameters by setting the *random seed* parameter ("-s 123") so as to be able to reproduce each run.

The results in [15] have shown that the 2-node transformation is in general more effective than the 3-node one, and preliminary experiments on the new real-world instances we considered confirmed this indication. Thus, the computational results reported in the next section are given only for the 2-node transformation for which parameter M has been set to 100,000.

Although a fine tuning of the Concorde parameters is out of the scope of this paper, we also analyzed the code sensitivity to the "chunk size" parameter, which controls the implementation of the *local cuts* paradigm used for separation (see, Applegate, Bixby, Chvátal and Cook [2] and Naddef [25] for details). In particular, setting this size to 0 ("-C 0") disables the generation of the "local" cuts and lets Concorde behave as a pure STSP code, whereas option "-C 16" (the default) allows for the generation of additional cuts based on the "instance-specific" enumeration of partial solutions over vertex subsets of size up to 16. (Other values of the chunk size, namely 8, 24 and 32, have been preliminary considered but the results suggested that the default size turns out to be, in general, the most effective one.)

In our computational experiments we considered both versions of Concorde, with and without the generation of "local" cuts, so as to investigate the capability of the "local" cuts method to automatically generate cuts playing an important role for the ATSP instance to be solved. Thus, the computational results of the next section are given for both "-C 0" and "-C 16" settings.

4   Computational Experiments

The two branch-and-cut codes described in the previous sections have been computationally tested on a set of pretty large[2] real-world ATSP instances (see Table 1), namely:

- 5 instances provided by Balas [8];
- the 13 ATSP instances with $n \geq 100$ collected in the TSPLIB [28].
- 1 instance (ftv180) introduced in [15];
- 16 instances by Cirasella, Johnson, McGeoch and Zhang [12].

All instances have integer nonnegative costs, and are available, on request, from the authors. (For a detailed description of the instances the reader is referred to [12], and [15].) For each instance we report in Table 1 the name (Name), the size (n), the optimal (or best known) solution value (OptVal), and the source of the instance (source).

All tests have been executed on a Digital Alpha 533 MHz with 512 MB of RAM memory under the Unix Operating System, with Cplex 6.5.3 as LP solver. In Table 2, we report the percentage gaps corresponding to the lower bound at the root node (Root), the final lower bound (fLB), and the final upper bound (fUB), all computed with respect to the optimal (or best known) solution value. Moreover, the number of nodes of the search tree (Nodes), and the computing time in seconds (Time) are given.

Code comparison. Table 2 reports a comparison among the two versions of Concorde corresponding to the chunk size equal to 0 and 16, respectively, and the branch-and-cut code implementing the algorithm by Fischetti and Toth described in Section 2. This latter code is enhanced, for the branching scheme, by

---

[2] Instances with less than 100 vertices have not been considered for space constraints.

Table 1. Real-world ATSP instances.

Name	n	OptVal	source	Name	n	OptVal	source
balas108	108	152	[8]	atex8	600	39982	[12]
balas120	120	286	[8]	big702	702	79081	[12]
balas160	160	397	[8]	code198	198	4541	[12]
balas200	200	403	[8]	code253	253	106957	[12]
ftv100	101	1788	TSPLIB	dc112	112	11109	[12]
ftv110	111	1958	TSPLIB	dc126	126	123235	[12]
ftv120	121	2166	TSPLIB	dc134	134	5612	[12]
ftv130	131	2307	TSPLIB	dc176	176	8587	[12]
ftv140	141	2420	TSPLIB	dc188	188	10225	[12]
ftv150	151	2611	TSPLIB	dc563	563	25951	[12]
ftv160	161	2683	TSPLIB	dc849	849	37476	[12]
ftv170	171	2755	TSPLIB	dc895	895	107699	[12]
ftv180	181	2918	[15]	dc932	932	479900	[12]
kro124p	100	36230	TSPLIB	td100.1	101	268636	[12]
rbg323	323	1326	TSPLIB	td316.10	317	691502	[12]
rbg358	358	1163	TSPLIB	td1000.20	1001	1242183	[12]
rbg403	403	2465	TSPLIB				
rbg443	443	2720	TSPLIB				

the "Fractionality Persistency" mechanism proposed by Fischetti, Lodi, Martello and Toth [14] and is called FT-b&c in the sequel and "FT-b&c + FP" in Table 2. As to time limit, we imposed 10,000 CPU seconds for all tests.

In its pure STSP version ("-C 0"), the Concorde code obtains a root-node lower bound which is dominated by the FT-b&c one, thus showing the effectiveness of addressing the ATSP in its original (directed) version. Of course, one can expect to improve the performance of FT-b&c by exploiting additional classes of ATSP-specific cuts, e.g., lifted cycle inequalities [10]. As to Concorde, we observe that the use of the "local" cuts leads to a considerable improvement of the root-node lower bound which becomes generally better than that of FT-b&c. Not surprisingly, this improvement appears more substantial than in the case of pure STSP instances. In our view, this is again an indication of the importance of exploiting the structure of the original asymmetric problem, which results into a very special structure of its STSP counterpart which is not captured adequately by the usual classes of STSP cuts (comb, clique tree inequalities, etc.).

Both FT-b&c and Concorde ("-C 16" version) turn out to be quite effective, and only fail in solving to optimality (within the time limit) 3 hard instances of large size. Specifically, the first code is considerably faster than the second one (with the only exception of instance ftv180), though it often requires more branching nodes. We believe this is mainly due to the faster (ATSP-specific) separation and pricing tools used in FT-b&c.

The Concorde implementation proved to be very robust for hard instances of large size (the final gap for the three unsolved instances being smaller than the one of FT-b&c), as it has been designed and engineered to address very large STSP instances.

Table 2. Comparison of branch-and-cut codes. Time limit of 10,000 seconds.

Name	(2-node) Concorde -s 123 -C 0 %Gap Root	fLB	fUB	Nodes	Time	Concorde -s 123 -C 16 %Gap Root	fLB	fUB	Nodes	Time	FT-b&c + FP %Gap Root	fLB	fUB	Nodes	Time
balas108	2.63	–	–	1023	1269.9	2.63	–	–	423	1416.0	1.97	–	–	267	89.0
balas120	2.10	0.00	1.40	2849	10007.3	1.05	–	–	755	7186.9	1.05	–	–	1339	1276.3
balas160	2.02	0.25	4.03	2165	10007.1	1.26	–	–	739	7848.0	1.26	–	–	737	671.1
balas200	2.48	0.74	5.96	1955	10008.0	0.74	–	–	239	2294.2	1.24	–	–	1495	1712.8
ftv100	0.73	–	–	9	9.5	0.00	–	–	1	12.6	0.39	–	–	21	2.2
ftv110	0.97	–	–	17	17.7	0.05	–	–	3	25.6	0.77	–	–	77	7.4
ftv120	1.62	–	–	69	45.0	0.28	–	–	7	54.4	0.97	–	–	123	13.1
ftv130	0.78	–	–	43	24.0	0.00	–	–	1	16.6	0.35	–	–	7	1.6
ftv140	0.45	–	–	5	7.2	0.00	–	–	3	25.6	0.25	–	–	9	2.1
ftv150	0.73	–	–	13	15.9	0.00	–	–	5	27.0	0.27	–	–	21	2.6
ftv160	0.93	–	–	29	70.0	0.30	–	–	7	55.7	0.67	–	–	17	3.8
ftv170	1.27	–	–	19	36.5	0.40	–	–	3	41.9	0.87	–	–	15	4.1
ftv180	1.58	–	–	91	204.0	0.69	–	–	29	236.2	1.20	–	–	939	366.0
kro124p	0.46	–	–	21	23.9	0.00	–	–	1	9.9	0.04	–	–	3	1.0
rbg323	0.00	–	–	7	34.4	0.00	–	–	3	23.9	0.00	–	–	1	0.4
rbg358	0.00	–	–	3	22.0	0.00	–	–	3	29.3	0.00	–	–	1	0.5
rbg403	0.00	–	–	1	19.7	0.00	–	–	5	49.3	0.00	–	–	1	1.3
rbg443	0.00	–	–	1	20.7	0.00	–	–	3	34.5	0.00	–	–	1	1.4
atex8	1.16	0.95	7.39	595	10080.8	1.09	0.65	7.62	143	10188.0	1.01	0.99	39.97	919	10000.5
big702	0.00	–	–	7	70.4	0.00	–	–	5	67.6	0.00	–	–	1	1.7
code198	0.00	–	–	1	23.3	0.00	–	–	1	29.6	0.00	–	–	1	0.4
code253	0.74	–	–	3	26.3	0.00	–	–	1	69.8	0.00	–	–	1	3.2
dc112	0.02	–	–	49	77.6	0.00	–	–	9	63.3	0.00	–	–	10	2.7
dc126	0.00	–	–	11	26.9	0.00	–	–	1	47.6	0.00	–	–	10	4.4
dc134	0.00	–	–	27	44.7	0.00	–	–	3	25.1	0.00	–	–	7	2.4
dc176	0.01	–	–	27	83.6	0.01	–	–	11	152.4	0.01	–	–	6	7.2
dc188	0.02	–	–	25	91.3	0.00	–	–	7	72.1	0.01	–	–	6	11.2
dc563	0.01	–	–	175	2615.6	0.00	–	–	35	827.7	0.00	–	–	69	343.5
dc849	0.00	–	–	45	1633.4	0.00	–	–	15	713.8	0.00	–	–	45	302.9
dc895	0.03	0.00	0.12	201	10298.1	0.01	0.00	0.07	89	10871.2	0.01	0.00	0.59	298	10008.4
dc932	0.01	0.00	0.03	237	10238.1	0.01	0.00	0.06	115	10457.4	0.01	0.01	0.27	211	10000.0
td100.1	0.00	–	–	1	2.0	0.00	–	–	1	2.4	0.00	–	–	1	0.0
td316.10	0.00	–	–	1	10.7	0.00	–	–	5	29.1	0.00	–	–	1	0.2
td1000.20	0.00	–	–	1	57.0	0.00	–	–	1	56.6	0.00	–	–	1	2.3

5   Conclusions

We considered a set of 35 real-world ATSP instances (with $n \geq 100$), and we computationally tested the effectiveness of branch-and-cut approaches designed either for the *asymmetric* version of the problem (Fischetti and Toth [17], FT-b&c code) or for its *symmetric* special case (Applegate, Bixby, Chvátal and Cook [2], Concorde code).

The fact that the performance of FT-b&c is generally better than that of the very sophisticated Concorde code (and considerably better than that of the pure STSP Concorde "-C 0") indicates the effectiveness of exploiting the ATSP-specific separation procedures. This suggests that enriching the Concorde arsenal of STSP separation tools by means of ATSP-specific separation procedures would be the road to go for the effective solution of hard ATSP instances.

Acknowledgments.   Work   supported   by   Ministero   dell'Istruzione, dell'Università e della Ricerca (M.I.U.R.) and by Consiglio Nazionale delle Ricerche (C.N.R.), Italy.

References

1. D. Applegate, R.E. Bixby, V. Chvátal, and W. Cook. Concorde - a code for solving traveling salesman problems. 12/15/1999 Release.
   http://www.keck.caam.rice.edu/concorde.html.
2. D. Applegate, R.E. Bixby, V. Chvátal, and W. Cook. On the solution of traveling salesman problems. *Documenta Mathematica*, Extra Volume ICM III:645–656, 1998.
3. N. Ascheuer. *Hamiltonian Path Problems in the On-line Optimization of Flexible Manufacturing Systems*. PhD thesis, Technische Universität Berlin, Germany, 1995.
4. N. Ascheuer, M. Fischetti, and M. Grötschel. A polyhedral study of the asymmetric travelling salesman problem with time windows. *Networks*, 36:69–79, 2000.
5. N. Ascheuer, M. Fischetti, and M. Grötschel. Solving the asymmetric travelling salesman problem with time windows by branch-and-cut. *Mathematical Programming, Ser. A*, 90:475–506, 2001.
6. N. Ascheuer, M. Jünger, and G. Reinelt. A branch & cut algorithm for the asymmetric traveling salesman problem with precedence constraints. *Computational Optimization and Applications*, 17:61–84, 2000.
7. E. Balas. The asymmetric assignment problem and some new facets of the traveling salesman polytope on a directed graph. *SIAM Journal on Discrete Mathematics*, 2:425–451, 1989.
8. E. Balas. Personal communication, 2000.
9. E. Balas and M. Fischetti. A lifting procedure for the asymmetric traveling salesman polytope and a large new class of facets. *Mathematical Programming, Ser. A*, 58:325–352, 1993.
10. E. Balas and M. Fischetti. Polyhedral theory for the asymmetric traveling salesman problem. In G. Gutin and A. Punnen, editors, *The Traveling Salesman Problem and its Variations*, pages 117–168. Kluwer Academic Publishers, 2002.

11. G. Carpaneto, M. Dell'Amico, and P. Toth. Algorithm CDT: a subroutine for the exact solution of large-scale asymmetric traveling salesman problems. *ACM Transactions on Mathematical Software*, 21:410–415, 1995.

12. J. Cirasella, D.S. Johnson, L.A. McGeoch, and W. Zhang. The asymmetric traveling salesman problem: Algorithms, instance generators, and tests. In A.L. Buchsbaum and J. Snoeyink, editors, *Proceedings of ALENEX'01*, volume 2153 of *Lecture Notes in Computer Science*, pages 32–59. Springer-Verlag, Heidelberg, 2001.

13. M. Fischetti. Facets of the asymmetric traveling salesman polytope. *Mathematics of Operations Research*, 16:42–56, 1991.

14. M. Fischetti, A. Lodi, S. Martello, and P. Toth. A polyhedral approach to simplified crew scheduling and vehicle scheduling problems. *Management Science*, 47:833–850, 2001.

15. M. Fischetti, A. Lodi, and P. Toth. Exact methods for the asymmetric traveling salesman problem. In G. Gutin and A. Punnen, editors, *The Traveling Salesman Problem and its Variations*, pages 169–205. Kluwer Academic Publishers, 2002.

16. M. Fischetti and P. Toth. An additive bounding procedure for the asymmetric travelling salesman problem. *Mathematical Programming, Ser. A*, 53:173–197, 1992.

17. M. Fischetti and P. Toth. A polyhedral approach to the asymmetric traveling salesman problem. *Management Science*, 43:1520–1536, 1997.

18. A.M.H. Gerards and A. Schrijver. Matrices with the Edmonds-Johnson property. *Combinatorica*, 6:365–379, 1986.

19. M. Grötschel and M.W. Padberg. Polyhedral theory. In E.L. Lawler, J.K. Lenstra, A.H.G. Rinnooy Kan, and Shmoys D.B. eds., editors, *The Traveling Salesman Problem: A Guided Tour of Combinatorial Optimization*, pages 251–305. Wiley, Chichester, 1985.

20. R. Jonker and T. Volgenant. Transforming asymmetric into symmetric traveling salesman problems. *Operations Research Letters*, 2:161–163, 1983.

21. M. Jünger, G. Reinelt, and G. Rinaldi. The traveling salesman problem. In M. Ball, T.L. Magnanti, C.L. Monma, and G. Nemhauser, editors, *Network Models*, volume 7 of *Handbooks in Operations Research and Management Science*, pages 255–330. North Holland, Amsterdam, 1995.

22. R.M. Karp. Reducibility among combinatorial problems. In R.E. Miller and J.W. Thatcher, editors, *Complexity of Computer Computations*, pages 85–103. Plenum Press, New York, 1972.

23. A. Löbel. Vehicle scheduling in public transit and Lagrangean pricing. *Management Science*, 44:1637–1649, 1998.

24. S. Martello. An enumerative algorithm for finding Hamiltonian circuits in a directed graph. *ACM Transactions on Mathematical Software*, 9:131–138, 1983.

25. D. Naddef. Polyhedral theory and branch-and-cut algorithms for the symmetric TSP. In G. Gutin and A. Punnen, editors, *The Traveling Salesman Problem and its Variations*, pages 29–116. Kluwer Academic Publishers, 2002.

26. M.W. Padberg and G. Rinaldi. An efficient algorithm for the minimum capacity cut problem. *Mathematical Programming, Ser. A*, 47:19–36, 1990.

27. M.W. Padberg and G. Rinaldi. Facet identification for the symmetric traveling salesman polytope. *Mathematical Programming, Ser. A*, 47:219–257, 1990.

28. G. Reinelt. TSPLIB - a traveling salesman problem library. *ORSA Journal on Computing*, 3:376–384, 1991. http://www.crpc.rice.edu/softlib/tsplib/.

The Bundle Method for Hard Combinatorial Optimization Problems

Gerald Gruber[1] and Franz Rendl[2]

[1] Carinthia Tech Institute, School of Geoinformation, A–9524 Villach, Austria
g.gruber@cti.ac.at
[2] University of Klagenfurt, Department of Mathematics, A–9020 Klagenfurt, Austria
franz.rendl@uni-klu.ac.at

Abstract. Solving the well known relaxations for large scale combinatorial optimization problems directly is out of reach. We use Lagrangian relaxations and solve it with the bundle method. The cutting plane model at each iteration which approximates the original problem can be kept moderately small and we can solve it very quickly. We report successful numerical results for approximating maximum cut.

1 Introduction

In combinatorial optimization many problems are very hard to solve. They have different tractable relaxations, e.g., linear programming relaxations which are based on the study of the convex hull of its integer solutions. Unfortunately for most problems no practical efficient method is known to optimize exactly over these relaxations and therefore usually a partial description by linear inequalities is used. Another valid refinement is semidefinite programming where striking advances have been achieved in the nineties ([6,23]). Nevertheless there is one drawback in common: For large scale problems it still remains hard to solve these modified relaxations directly (high computational effort, large memory requirements, etc.).

Recently, great efforts have been made to tackle this problem and several papers were published. For instance, Benson, Ye and Zhang [4] present a dual-scaling interior point algorithm and show how it exploits the structure and sparsity of some large scale problems. Helmberg and Rendl propose the spectral bundle method [12]. They also exploit the sparsity structure of the problems. In fact, they consider eigenvalue optimization problems arising from semidefinite programming problems with constant trace and present a method which leads to reasonable bounds to the optimal solution of large problems. Barahona and Anbil propose the volume algorithm [2] for dealing with Lagrangian relaxations of combinatorial optimization problems. The volume algorithm is an extension of the subgradient method which is an iterative technique where iterates are updated using a current subgradient and a carefully chosen step size. In [2] very sucessful experiments with linear programs coming from combinatorial problems are presented.

M. Jünger et al. (Eds.): Combinatorial Optimization (Edmonds Festschrift), LNCS 2570, pp. 78–88, 2003.

We follow the idea of using Lagrangian relaxation but solve it with the bundle method instead of simple subgradient methods. Our main contribution is that we allow inequalities as constraints. We show how they can be handled and present promising numerical results.

This paper is organized as follows: In section 2 the problem in general form is stated. In this section we also show how the maximum cut problem fits into this general setting. In section 3 we review Lagrangian relaxation, the bundle method and the bundle algorithm. In section 4 we present first numerical experiments for the maximum cut problem on different randomly generated graphs. We conclude with final remarks.

2   The Problem in General Form

We consider the problem of maximizing a concave function $c : \mathbb{R}^n \to \mathbb{R}$ subject to finitely many linear constraints. In particular, we deal with the problem of the following form.

Given $c : \mathbb{R}^n \to \mathbb{R}$, $A \in \mathbb{R}^{m \times n}$, $b \in \mathbb{R}^m$ and a convex set X of "nice structure", solve

$$(\mathbf{P}) \qquad \max c(x)$$
$$\text{s.t. } x \in X$$
$$Ax \le b.$$

We do not specify the set X exactly, it could be a simple polyhedron or some other simple structure. By "nice structure" we mean that optimizing over $x \in X$ can be done efficiently, while maintaining explicitly the additional inequality constraints $Ax \le b$ may be computationally prohibitive.

Solving $(\mathbf{P})$ directly is out of reach. The question is: How can we deal with $(\mathbf{P})$ better, how can we select important inequalities of $Ax \le b$ to get an approximation of the original problem, as good as possible? In this paper we present a method which leads to such an approximation in reasonable time.

2.1   The Maximum Cut Problem

Many combinatorial problems fit exactly into this framework. To give a specific example, we consider the maximum cut problem (max-cut). Given an edge-weighted undirected graph $G = (V, E)$ with vertex set $V = \{1, \dots, n\}$ and edge set $E \subseteq \{(ij) : i, j \in V, i < j\}$ the max-cut problem is the problem of finding a partition $(S, V \backslash S)$ that maximizes the sum of the weight of the edges connecting S with $V \backslash S$.

The max-cut problem is known to be NP-hard, see Karp [16,17]. There are several classes of graphs for which the maximum cut problem can be solved in polynomial time. For instance for graphs with no long odd cycles, (see [9]), for planar graphs, (see [10,26]), or more generally for graphs with no K_5−minor (see [1]). For planar graphs or more generally graphs not contractible to K_5, it is known that the cut polytope coincides with the metric polytope, hence this relaxation would yield the optimal solution of max-cut.

Several relaxations have been studied and proposed in the literature. (see e.g. [1,3,15] for LP-based relaxations, e.g. [25,27] for eigenvalue based relaxations and e.g. [6,8] for semidefinite relaxations.) More recently, the following well-known semidefinite relaxation of max-cut was introduced, see e.g. [6,8,24,5].

$$\max \operatorname{trace}(LX) \text{ s.t. } \operatorname{diag}(X) = e, \ X \succeq 0. \tag{1}$$

The quality of this bound was analyzed by Goemans and Williamson [8]. Assuming nonnegative weights they have shown that the ratio between the optimal value of max-cut and the semidefinite upper bound is at least .878. For completeness we mention that it is NP-complete to approximate the maximum cut problem with a factor closer than .9412, see [14].

This model can be strenghtened by imposing in addition that X is contained in the metric polytope ($X \in MET$). By definition,

$$X \in MET \ :\Leftrightarrow \ \begin{cases} X_{ii} = 1 & \text{for } i = 1, \dots, n, \\ X_{ij} - X_{ik} - X_{jk} \geq -1 \text{ for } 1 \leq i, j, k \leq n, \\ X_{ij} + X_{ik} + X_{jk} \geq -1 \text{ for } 1 \leq i, j, k \leq n. \end{cases} \tag{2}$$

Poljak and Rendl [28] were the first who proposed to strengthen the semidefinite relaxation by including the triangle inequalities. Computational results are presented for instance in [13,29].

The semidefinite model is of the form described above. ($Ax \leq b$ corresponds to $X \in MET$, while the easy subproblem is a semidefinite program.)

3   The Bundle Method

In this section we shall derive the Lagrangian dual associated to **(P)** which leads to our starting point. We use the following corollary, which is an easy consequence of [30, Theorem 37.3].

Corollary 1. *Let X and Y be non–empty closed convex sets in $\mathbb{R}^m$ and $\mathbb{R}^n$, respectively, and let f be a continous finite concave - convex function on $X \times Y$. If either X or Y is bounded, one has*

$$\inf_{y \in Y} \sup_{x \in X} f(x, y) = \sup_{x \in X} \inf_{y \in Y} f(x, y).$$

(In general for a non-empty product set $X \times Y$ and $f : X \times Y \mapsto [-\infty, +\infty]$ one has: $\inf_{x \in X} \sup_{y \in Y} f(x, y) \geq \sup_{y \in Y} \inf_{x \in X} f(x, y)$, see [30, Lemma 36.1])

Let $\gamma \in \mathbb{R}^m, \gamma \geq 0$ denote the Lagrangian multiplier for $b - Ax \geq 0$, then

$$\max c(x) \text{ s.t. } x \in X, Ax \leq b = \max_{x \in X} \min_{\gamma \geq 0} c(x) + \gamma^T (b - Ax)$$

$$= \min_{\gamma \geq 0} \max_{x \in X} c(x) + \gamma^T (b - Ax)$$

$$\leq f(\gamma), \forall \gamma \geq 0$$

where $f(\gamma) := \max_{x \in X} c(x) + \gamma^T (b - Ax)$.

Now consider the translated problem $\min_{\gamma \geq 0} f(\gamma)$. It is still too difficult to solve it directly.

As abbreviation for $c(x) + \gamma^T (b - Ax)$ we use $L(\gamma, x)$.

Given γ, we set

$$L(\gamma, x(\gamma)) := f(\gamma) = \max_{x \in X} L(\gamma, x).$$

Fact 1. *Suppose that $L(\gamma, x)$ is affine linear in γ and X is non-empty. Then*

$$f(\gamma) := \max_{x \in X} L(\gamma, x) \text{ is convex.}$$

Proof. Let $\alpha \in [0, 1]$ and choose $\gamma_i \geq 0, i = 1, 2$. We set $\gamma_3 := \alpha\gamma_1 + (1 - \alpha)\gamma_2$. By definition $(\gamma_i, x(\gamma_i))$ maximizes $L(\gamma_i, x)$, $i = 1, 2$ For this reason the following inequality holds.

$$L(\gamma_1, x(\gamma_3)) \leq \alpha L(\gamma_1, x(\gamma_1)) + (1 - \alpha)L(\gamma_2, x(\gamma_2)).$$

$\square$

Moreover, $f(\gamma)$ is non-smooth and therefore classical methods from analysis for minimizing $f(\gamma)$ will fail completely. A feasible method for handling optimization problems with non-differentiable convex cost functions is the bundle concept, see e.g. [31,19,21]. Bundle methods were first proposed by Lemarechal [21]. The method has developed over the time based on the papers of Kiwiel [18] and Lemarechal [22]. We follow the approach of Helmberg, Kiwiel, Rendl [11] and present the details in the following subsection.

3.1   Minimizing $f(\gamma)$

Given γ_0 we compute $f(\gamma_0) = L(\gamma_0, x_0)$. Clearly, $L(\gamma, x_0) \leq \max_{x \in X} L(\gamma, x) = f(\gamma)$. Hence, for given $\gamma \geq 0$ $L(\gamma, x_0)$ is a minorant of $f(\gamma)$ which approximates $f(\gamma)$ near γ_0.

Now we assume that $f(\gamma)$ is evaluated at $\gamma_0, \ldots, \gamma_k$. Let

$$\hat{f}^k(\gamma) = \max\{L(\gamma, x_j) : 0 \leq j \leq k\}.$$

Note, $\hat{f}^k(\gamma)$ need not be bounded from below. To make sure that $\hat{f}^k(\gamma)$ is always bounded from below we change $\hat{f}^k(\gamma)$ by adding a quadratic regularization term resulting in

$$f^k(\gamma) = \hat{f}^k(\gamma) + \frac{1}{2t}||\gamma - \bar{\gamma}||^2, \ t \geq 0 \text{ fixed.}$$

Here $\bar{\gamma}$ denotes the current best approximation of the minimizer of $f(\gamma)$. ($f^k(\gamma)$ has a minimum since $\hat{f}^k(\gamma)$ decreases linearly and $||\gamma - \bar{\gamma}||^2$ increases quadratically.) The fixed parameter t controls the step size, large steps will be punished. The choice of t turns out to be very critical. Capable estimates for t are given in [20].

We next show how we use Lagrangian methods for solving the subproblem $\min_{\gamma \geq 0} f^k(\gamma)$.

Let $X^k = (x_1, \ldots, x_k)$, $c(X^k) = (c(x_1), \ldots, c(x_k))$ and $\gamma_{k+1} \geq 0$ such that $f^k(\gamma_{k+1}) = \min\limits_{\gamma \geq 0} f^k(\gamma)$.

Hence

$$f^k(\gamma_{k+1})$$

$$= \min_{\gamma \geq 0} \max_{0 \leq j \leq k} L(\gamma, x_j) + \frac{1}{2t}||\gamma - \bar{\gamma}||^2$$

$$= \min_{\gamma} \max_{0 \leq j \leq k} \max_{\eta \geq 0} L(\gamma, x_j) + \frac{1}{2t}||\gamma - \bar{\gamma}||^2 - \gamma^T \eta$$

$$= \max_{\eta \geq 0} \max_{\lambda \geq 0, \sum_j \lambda_j = 1} \min_{\gamma} \sum_j \lambda_j L(\gamma, x_j) + \frac{1}{2t}||\gamma - \bar{\gamma}||^2 - \gamma^T \eta$$

Now observe that the inner minimization is unconstrained, $\frac{\partial}{\partial \gamma}(\cdot) = 0$ is necessary and sufficient for optimality.

$$\frac{\partial}{\partial \gamma}(\cdot) = 0 \Leftrightarrow \sum_{j=1}^{k} \lambda_j \underbrace{(b - Ax_j)}_{:=g_j} + \frac{1}{t}(\gamma - \bar{\gamma}) - \eta = 0$$

From this it easily follows that $\gamma = \bar{\gamma} + t(\eta - G\lambda)$ where $G := (g_0, \ldots, g_k) \in \mathbb{R}^{m \times k}$. Using this our problem translates into

$$\max_{\eta \geq 0} \max_{\lambda \geq 0, \sum_j \lambda_j = 1} \sum_j \lambda_j c(x_j) + \langle \bar{\gamma} + t(\eta - G\lambda), G\lambda \rangle + \frac{1}{2t}||t(\eta - G\lambda)||^2$$

$$- \langle \bar{\gamma} + t(\eta - G\lambda), \eta \rangle$$

$$= \max_{\eta \geq 0} \max_{\lambda \geq 0, \sum_j \lambda_j = 1} \sum_j \lambda_j c(x_j) - \langle \bar{\gamma} + t(\eta - G\lambda), \eta - G\lambda \rangle$$

$$+ \frac{t}{2}\langle \eta - G\lambda, \eta - G\lambda \rangle$$

$$= \max_{\eta \geq 0} \max_{\lambda \geq 0, \sum_j \lambda_j = 1} \sum_j \lambda_j c(x_j) + \langle \bar{\gamma}, G\lambda \rangle - \langle \bar{\gamma}, \eta \rangle - \frac{t}{2}\langle \eta - G\lambda, \eta - G\lambda \rangle$$

$$= \max_{\eta \geq 0} \max_{\lambda \geq 0, \sum_j \lambda_j = 1} -\frac{t}{2}\langle \eta - G\lambda, \eta - G\lambda \rangle - \langle \bar{\gamma}, \eta \rangle - \langle \lambda, \beta \rangle$$

where $\beta := -c(X^k)^T - G^T \bar{\gamma}$.

Next we show, how to solve this problem efficiently. First we consider the problem for fixed $\eta \geq 0$ and maximize over $\lambda \geq 0$. Then we fix the optimal λ and maximize over η. This process is iterated several times.

Suppose $\bar{\eta} \geq 0$ is fixed, then the inner maximization is

$$\bar{\lambda} := \text{argmax}_{\lambda \geq 0, \sum_j \lambda_j = 1} -\frac{t}{2}\lambda^T G^T G\lambda + \lambda^T(-\beta + tG^T\bar{\eta}) \underbrace{-\frac{t}{2}\eta^T\eta - \eta^T\bar{\gamma}}_{const}.$$

In the second step we keep $\bar{\lambda}$ fixed and have to solve

$$\bar{\eta} := \text{argmax}_{\eta \geq 0} -\frac{t}{2}\eta^T\eta + t\bar{\lambda}^T G^T\eta - \bar{\gamma}^T\eta - \bar{\lambda}^T\beta - \frac{t}{2}\bar{\lambda}^T G^T G\bar{\lambda}.$$

Consider the problem $\max_{u \geq 0} h(u)$ where $h(u) := -\frac{t}{2}u^2 + au, a \in \mathbb{R}$. Obviously

$$\frac{\partial}{\partial u}(h(u)) = 0 \Leftrightarrow \text{argmax}(h(u)) = \begin{cases} \frac{a}{t} & \text{if } \frac{a}{t} \geq 0 \\ 0 & \text{otherwise.} \end{cases}$$

Now it is easy to see that

$$\bar{\eta}_i = \begin{cases} \frac{1}{t}(-\gamma_{k,i} + t(G\lambda)_i) & \text{if } (-\gamma_{k,i} + t(G\lambda)_i) \geq 0 \\ 0 & \text{otherwise.} \end{cases}$$

This "zig zag steps" yield a pair $(\bar{\eta}, \bar{\lambda})$, which we use for computing the step direction $d := t(\bar{\eta} - G\bar{\lambda})$.

Hence

$$(\gamma_k + d)_i = \begin{cases} 0 & \text{if } (-\gamma_{k,i} + t(G\lambda)_i) \geq 0 \\ \gamma_{k,i} - t(G\lambda)_i & \text{otherwise.} \end{cases}$$

We use $\gamma_{k+1} := \gamma_k + d$ as new trial point.

3.2 The Algorithm

In this section we discuss the essential ingredients of our implementation. Our initial relaxation is the original **(P)** without inequality constraints. The optimal solution of it gives us the first separation point x_s, which in general violates a bunch of inequalities of $Ax \leq b$. We only choose the most violated ones and for those we perform the following steps.

(i)   construct the model in the k-th step:

$$f^k(\gamma) := \max_{0 \leq i \leq k} L(\gamma, x_i) + \frac{1}{2t}\|\gamma - \bar{\gamma}\|^2$$

(ii) compute γ_{k+1} the minimizer of $f^k(\gamma)$

(iii) compute x_{k+1} such that $f(\gamma_{k+1}) = L(\gamma_{k+1}, x_{k+1})$

(iv) compute the current best approximation of the minimizer of $f(\gamma)$

$$\bar{\gamma} := \begin{cases} \gamma_{k+1} & \text{if } \gamma_{k+1} \text{ better than } \bar{\gamma} \\ \bar{\gamma} & \text{otherwise.} \end{cases}$$

The minimizer in step 2 is computed using the "zig zag idea" proposed above. The optimal pair $(\bar{\eta}, \bar{\lambda})$ gives us some information, namely

(i) $\bar{\lambda} \approx 0$ tells us that the corresponding subgradient is inactive.
(ii) $\bar{\eta} \approx 0$ tells us that the corresponding inequality in the current set of hard constraints is inactive.

In other words we can control the bundle size and we are able to keep only important inequalities from $Ax \leq b$.

In the $k'th$ turn of the steps above we update x_s to a certain convex combination of x_s and x_{k+1}. Our next separation point is x_s.

Convergence follows from the traditional approach and therefore the proof is not embodied in the paper.

4   Computational Results

In section 2.1 we introduced max-cut. In this section we report computational experience for this NP-hard problem. As initial relaxation we use the semidefinite relaxation introduced in section 2.1. We strengthen it by adding triangle inequalities from (2). Our numerical experiments were carried out on several randomly generated graphs of various size and fixed density of 50 %. Our test data are available at http://www-sci.uni-klu.ac.at/math-or/home/index.htm. All our codes are implemented in Matlab using some interfaces in C. All computations are done on a Pentium II 400 Mhz computer.

It is the primary objective of this section to confront the bundle method with the interior point method. The computational results on our problem sets are reported in Table 1 and Table 2. Table 1 corresponds to the interior point approach and Table 2 to the bundle approach. Column n gives the number of vertices. The column labeled ib stands for "initial bound" and contains the optimal objective value from (1). The column headed ub stands for "upper bound" and shows the bound on the optimal value for max-cut which we achieved after some stopping criterion. The computation time using interior point methods increases drastically with the number of added inequalities and while solving the extended model a better part of them becomes inactive and can be dropped out from the model. Therefore we stopped the interior point code after twelve rounds of adding a limited number of violated inequalities. The bundle code was stopped after four rounds of adding inequalities. bc shows the best cut which we have found using a heuristic method. Column gap gives the relative error of the upper bound lb in % with respect to the values given in column bc. Computation times in minutes are given in Column $CPU-time$. The numbers given in act_{tri} represent the number of active inequality constraints of the final relaxation. Column $viol$ contains the average violation of triangle inequalities of the current primal solution.

The numbers in Table 1 and Table 2 indicate not only the limitation of pure interior point methods but also the potential of the bundle method. Problems of moderate size can be solved by both approaches. The qualitative behaviour is quite the same. On the other hand it is self-evident that large scale problems should not be tackeled by pure interior point methods. The small improvement of the bound is bought very dearly.

One important benefit of the bundle approach is that the cost for solving the basic relaxation in each iteration depends not on the number of inequalities. There is only a small change in the objective function. We point out this feature in Table 3. Again we compare our interior point method with our bundle method. The graphs for this experiment are those from above with 100, 200 nodes, respectively. *tri* denotes the number of triangles from MET in the current model. In Table 3 the numbers given in act_{tri} represent the number of active inequality constraints at the optimum from the current model. The columns assigned to the interior point results show how computation time increases as more and more triangle inequalities are included. This is not the case in this sheer enormity if we use our bundle code.

Table 1. Bounds on max-cut on random graphs using interior point methods, edge density = 50 %

n	ib	ub	bc/gap	$CPU - time[\min]$	act_{tri}	$viol$
100	8299.8	7436.1	6983/6.1	8.1	499	0.17
200	26904.2	24941.4	22675/9.1	56.1	1066	0.13
300	47661.6	44696.0	38648/13.5	116.5	1389	0.13
400	77056.1	73574.7	62714/14.8	213.1	1618	0.12
500	104518.7	100147.7	83491/16.6	337.7	1822	0.12

Table 2. Bounds on max-cut on random graphs using the bundle method, edge density = 50 %

n	ib	ub	bc/gap	$CPU - time[\min]$	act_{tri}	$viol$
100	8299.8	7520.9	6983/7.2	2.4	805	0.036
200	26904.2	25093.6	22675/9.6	10.1	1551	0.026
300	47661.6	45081.3	38648/14.3	26.3	1636	0.026
400	77056.1	73971.3	62714/15.2	55.5	1957	0.024
500	104518.7	100818.9	83491/17.2	100.2	1946	0.024

Table 3. Comparing the behaviour of the interior point approach and the bundle method on a random weighted graph with 100 nodes and density = 50%

interior point method			bundle method		
ub	tri/act_{tri}	CPU [min]	ub	tri/act_{tri}	CPU [min]
8299.8	0/0	0.02	8299.8	0/0	0.02
7882.7	140/106	0.3	7874.9	500/243	0.5
7701.7	246/201	0.9	7671.9	743/484	1.1
7605.3	341/277	1.9	7574.9	984/686	1.7
7535.1	417/353	3.2	7514.3	928/764	2.3
7480.5	493/435	5.1	7473.2	1027/903	2.9
7435.6	575/502	8.0	7449.9	1108/992	3.6
7407.5	642/580	11.0	7440.2	1267/1129	4.3
7387.4	720/666	15.6	7434.1	1322/1171	5.0
7374.5	806/745	21.2	7425.6	1393/1259	5.7
7366.9	885/827	28.2	7420.3	1387/1271	6.4
7362.4	967/908	37.4	7415.1	1453/1335	7.2
7360.2	1048/993	47.8	7412.4	1501/1377	7.9
7359.6	1091/1067	59.3	7410.9	1544/1405	8.7
7359.6	1118/1105	75.5	7409.5	1561/1415	9.5

Table 4. Comparing the behaviour of the interior point approach and the bundle method on a random weighted graph with 200 nodes and density = 50%

interior point method			bundle method		
ub	tri/act_{tri}	CPU [min]	ub	tri/act_{tri}	CPU [min]
26904.2	0/0	0.1	26904.2	0/0	0.1
25982.9	280/193	1.9	26070.7	1000/409	2.2
25521.7	473/380	5.0	25510.3	1409/824	4.8
25297.6	660/556	10.9	25267.6	1824/1295	7.4
25145.9	836/744	21.3	25126.3	1640/1404	10.1
25036.6	1024/904	36.6	25040.2	1885/1670	12.8
24948.4	1184/1068	57.1	24999.1	2173/1932	15.6
24893.4	1348/1243	86.2	24969.9	2135/1979	18.4
24857.7	1523/1411	126.0	24946.0	2178/2013	21.3
24834.1	1691/1586	178.2	24932.0	2195/2073	24.2
24821.7	1866/1789	248.0	24924.0	2304/2173	27.0
			⋮		
24815.7	2069/2004	330.6	24820.6	2907/2669	44.7

5   Concluding Comments

Bundle methods are an effective solution approach for hard combinatorial optimization problems. In this paper we have presented an algorithm for approximating such problems. The most important feature is that the set of subgradients (the bundle) which is updated at each iteration can be kept moderately small. Therefore this approach allows to solve large scale problems very quickly.

In more detail we concentrated on max-cut. We have performed our numerical experiments on this problem. Our computational tests give rise to the following conclusions. Compared with semidefinite programming via interior point methods we obtained acceptable bounds in moderate time, even for large scale problems. The algorithm is easy to implement, but has slow convergence.

References

1. F. Barahona. The max-cut problem on graphs not contractible to K_5. *Operations Research Letters*, 2(3):107–111, 1983.
2. F. Barahona and R. Anbil. The volume algorithm: producing primal solutions with a subgradient method. *Mathematical Programming*, 87(3):385–399, 2000.
3. F. Barahona and A. R. Mahjoub. On the cut polytope. *Mathematical Programming*, 36:157–173, 1986.
4. S. J. Benson, Y. Ye, and X. Zhang. Solving large-scale sparse semidefinite programs for combinatorial optimization. *SIAM J. Optim.*, 10(2):443–461 (electronic), 2000.
5. C. Delorme and S. Poljak. Laplacian eigenvalues and the maximum cut problem. *Mathematical Programming*, 62(3):557–574, 1993.
6. M. X. Goemans. Semidefinite programming in combinatorial optimization. *Mathematical Programming*, 79(1–3):143–161, 1997.
7. M. X. Goemans and D. P. Williamson. .878-approximation algorithms for MAX CUT and MAX 2SAT. In *Proceedings of the Twenty-Sixth Annual ACM Symposium on the Theory of Computing*, pages 422–431, Montréal, Québec, Canada, 1994.
8. M. X. Goemans and D. P. Williamson. Improved approximation algorithms for maximum cut and satisfiability problems using semidefinite programming. *Journal of the ACM*, 42(6):1115–1145, 1995. preliminary version, see [7].
9. M. Grötschel and G. L. Nemhauser. A polynomial algorithm for the max-cut problem on graphs without long odd cycles. *Mathematical Programming*, 29:28–40, 1984.
10. F. Hadlock. Finding a maximum cut of a planar graph in polynomial time. *SIAM Journal on Computing*, 4:221–225, 1975.
11. C. Helmberg, K. C. Kiwiel, and F. Rendl. Incorporating inequality constraints in the spectral bundle method. In R. E. Bixby, E. A. Boyd, and R. Z. Ríos-Mercado, editors, *Integer Programming and Combinatorial Optimization*, volume 1412 of *Lecture Notes in Computer Science*, pages 423–435. Springer, 1998.
12. C. Helmberg and F. Rendl. A spectral bundle method for semidefinite programming. *SIAM Journal on Optimization*, 10(3):673–696, 2000.
13. C. Helmberg, F. Rendl, R. J. Vanderbei, and H. Wolkowicz. An interior point method for semidefinite programming. *SIAM Journal on Optimization*, 6(2):342–361, 1996.

14. J. Håstad. Some optimal inapproximability results. In *Proceedings of the 29th ACM Symposium on Theory of Computing (STOC)*, pages 1–10, 1997.

15. M. Jünger, F. Barahona, and G. Reinelt. Experiments in quadratic 0-1 programming. *Mathematical Programming*, 44(2):127–137, 1989.

16. R. M. Karp. Reducibility among combinatorial problems. In R. E. Miller and J. W. Thatcher, editors, *Complexity of Computer Computations*, pages 85–103, New York, 1972. Plenum Press.

17. R. M. Karp. On the computational complexity of combinatorial problems. *Networks*, 5(1):45–68, 1975.

18. K. C. Kiwiel. *Methods of Descent of Nondifferentiable Optimization*, volume 1133 of *Lecture Notes in Mathematics*. Springer, Berlin, 1985.

19. K. C. Kiwiel. Survey of bundle methods for nondifferentiable optimization. *Math. Appl. Jan. Ser.6*, 6:263–282, 1989.

20. K. C. Kiwiel. Proximity control in bundle methods for convex nondifferentiable minimization. *Mathematical Programming*, 46:105–122, 1990.

21. C. Lemarechal. Bundle methods in nonsmooth optimization. In Claude Lemarechal and Robert Mifflin, editors, *Proceedings of the IIASA Workshop, vol. 3, Nonsmooth Optimization, March 28-April 8*, pages 79–102. Pergamon Press, 1978, 1977.

22. C. Lemaréchal, A. Nemirovskii, and Yu. Nesterov. New variants of bundle methods. *Mathematical Programming*, 69:111–147, 1995.

23. L. Lovász. Semidefinite programs and combinatorial optimization. Lecture Notes, 1995.

24. M. Laurent, S. Poljak, F. Rendl. Connections between semidefinite relaxations of the max-cut and stable set problems. *Mathematical Programming*, 77(2):225–246, 1997.

25. B. Mohar and S. Poljak. Eigenvalues in combinatorial optimization. In *Combinatorial Graph-Theoretical Problems in Linear Algebra*, IMA Vol. 50. Springer-Verlag, 1993.

26. G. I. Orlova and Ya.G. Dorfman. Finding the maximum cut in a graph. *Engineering Cybernetics*, 10(3):502–506, 1972.

27. S. Poljak and F. Rendl. Node and edge relaxations for the max-cut problem. *Computing*, 52:123–127, 1994.

28. S. Poljak and F. Rendl. Nonpolyhedral relaxations of graph-bisection problems. *SIAM Journal on Optimization*, 5(3):467–487, 1995.

29. F. Rendl. Semidefinite programming and combinatorial optimization. *Applied Numerical Mathematics*, 29:255–281, 1999.

30. R. Tyrrell Rockafellar. *Convex Analysis*, volume 28 of *Princeton Mathematics Series*. Princeton University Press, Princeton, 1970.

31. H. Schramm and J. Zowe. A version of the bundle idea for minimizing a nonsmooth function: Conceptual idea, convergence analysis, numerical results. *SIAM J. Optim.*, 2:121–152, 1992.

The One-Commodity Pickup-and-Delivery Travelling Salesman Problem*

Hipólito Hernández-Pérez and Juan-José Salazar-González

DEIOC, Faculty of Mathematics, University of La Laguna
Av. Astrofísico Francisco Sánchez, s/n; 38271 La Laguna, Tenerife, Spain
{hhperez,jjsalaza}@ull.es

Abstract. This article deals with a new generalization of the well-known "Travelling Salesman Problem" (TSP) in which cities correspond to customers providing or requiring known amounts of a product, and the vehicle has a given capacity and is located in a special city called depot. Each customer and the depot must be visited exactly once by the vehicle serving the demands while minimizing the total travel distance. It is assumed that the product collected from pickup customers can be delivered to delivery customers. The new problem is called "one-commodity Pickup-and-Delivery TSP" (1-PDTSP). We introduce a 0-1 Integer Linear Programming model for the 1-PDTSP and describe a simple branch-and-cut procedure for finding an optimal solution. The proposal can be easily adapted for the classical "TSP with Pickup-and-Delivery" (PDTSP). To our knowledge, this is the first work on an exact method to solve the classical PDTSP. Preliminary computational experiments on a test-bed PDTSP instance from the literature show the good performances of our proposal.

1 Introduction

This article presents a branch-and-cut algorithm for a routing problem called *one-commodity Pickup-and-Delivery Travelling Salesman Problem* (1-PDTSP) and is closely related to the well-known *Travelling Salesman Problem* (TSP). A novelty of the 1-PDTSP compared to the TSP is that one specified city is considered to be a *depot*, while the others cities are associated to *customers* divided into two groups according to the type of required service. Each *delivery customer* requires a given amount of the product, while each *pickup customer* provides a given amount of the product. The product collected from a pickup customer can be supplied to a delivery customer, on the assumption that there will be no deterioration in the product. It is assumed that the vehicle has a fixed upper-limit *capacity*, starting and ending the route at the depot. The travel distance between each pair of locations is known. The 1-PDTSP calls for a minimum distance route for the vehicle satisfying the customer requirements without ever exceeding the vehicle capacity.

* Work partially supported by "Gobierno de Canarias" (PI2000/116) and by "Ministerio de Ciencia y Tecnología" (TIC2000-1750-C06-02), Spain.

M. Jünger et al. (Eds.): Combinatorial Optimization (Edmonds Festschrift), LNCS 2570, pp. 89–104, 2003.

This optimization problem finds applications in the transportation of a product from customer to customer when no unit of product has a precise origin and/or destination. This is the case, for example, when a bank company must move money between branch offices, some of them providing money and the others requiring money; the main office (i.e., the vehicle depot) provides or collects the remaining money. Another example occurs when milk must be distributed from farms to factories by a capacitated vehicle, assuming that each factory is only interested in receiving a stipulated demand of milk but not in the providers of this demand.

Clearly, the 1-PDTSP is an $\mathcal{NP}$-hard problem in the strong sense since it coincides with TSP when the vehicle capacity is large enough.

A close-related problem in the literature is the *Travelling Salesman Problem with Pickup and Delivery* (PDTSP). As in the 1-PDTSP, there are two types of customers, each with a given (unrestricted in sign) demand, and a vehicle with a given capacity stationed in a depot. Travel distances are also given. The main difference is that in the PDTSP the product collected from pickup customers is different from the product supplied to delivery customers. Therefore, the total amount of items collected from pickup customers must be delivered only to the depot, and there are other different items going from the depot to the delivery customers. In other words, the positive demand customers provide items of a first product that must transported to the depot, while the negative demand customers require items of a different second product that must be transported from the depot. Both products must share the vehicle capacity during the route. An application of PDTSP is the collection of empty bottles from customers for delivery to a warehouse and full bottles being delivered from the warehouse to the customers. An immediate difference when compared with 1-PDTSP is that PDTSP is only feasible when the vehicle capacity is at least the maximum of the total sum of the pickup demands and of the total sum of the delivery demands. This constraint is not required for the feasibility of the 1-PDTSP in which the vehicle capacity could even be equal to the biggest customer's demand.

Mosheiov [15] introduced the PDTSP and proposed applications and heuristic approaches. Anily and Mosheiov [3] and Gendreau, Laporte and Vigo [9] presented approximation algorithms for the PDTSP. Anily and Hassin [2] introduced the *Swapping Problem*, the particular case of 1-PDTSP in which the customer requirements and the vehicle capacity are identical, and presented a polynomial approximation algorithm.

Another related problem is the *TSP with Backhauls-and-Linehauls*, where there is the additional constraint that delivery customers must be visited before any pickup customer. In the *Dial-a-Ride TSP* there is a one-to-one correspondence between pickup customers and delivery customers, and each delivery customer must be visited only after the corresponding pickup customer has been visited. When there is no vehicle capacity, the Dial-a-Ride TSP is also known as *Stacker Crane Problem* and it is a particular case of the *TSP with Precedence Constraints*. See Savelsbergh and Sol [18] for references and other variants including time-windows, several vehicles, etc. The 1-PDTSP is also related to the

Capacitated Vehicle Routing Problem (CVRP), where a homogeneous capacitated vehicle fleet located in a depot must collect a product from a set of pickup customers. CVRP combines structure from the TSP and from the Bin Packing Problem, two well-known and quite different combinatorial problems. See Toth and Vigo [19] for a recent survey on this routing problem.

To our knowledge this is the first article introducing and solving the 1-PDTSP. Section 2 shows that even finding a feasible solution of this combinatorial problem is $\mathcal{NP}$-hard in the strong sense and Section 3 relates PDTSP to 1-PDTSP. Section 4 presents 0-1 integer linear programming models for the asymmetrical and symmetrical cases, and Section 5 describes a branch-and-cut algorithm for finding an optimal solution. Some preliminary computational results on a test-bed PDTSP instance from Mosheiov [15] are presented in Section 6 to analyze the performance of our proposal solving both 1-PDTSP and classical PDTSP instances.

2   Computational Complexity

As mentioned before, 1-PDTSP is an $\mathcal{NP}$-hard optimization problem, since it reduces to the TSP when Q is large enough (e.g. bigger than the sum of the delivery demands and than the sum of pickup demands). This section shows that even finding a feasible 1-PDTSP solution (not necessarily an optimal one) can be a very complex task.

Indeed, let us consider the so-called *3-partitioning problem*, defined as follows. Let us consider a positive integer number P and a set I of $3m$ items. Each item i is associated with a positive integer number p_i such that $P/4 \leq p_i \leq P/2$ and $\sum_{i \in I} p_i = mP$. Can the set I be partitioned into m disjoint subsets $I_1, \ldots, I_m$ such that $\sum_{i \in I_k} p_i = P$ for all $k \in \{1, \ldots, m\}$?

Clearly, this problem coincides with the problem of checking if there is (or not) a feasible solution of the following 1-PDTSP instance. Consider a pickup customer with demand $q_i = p_i$ for each item $i \in I$, m delivery customers with demand $-P$, and a vehicle with capacity equal to P. Since we are interested only in a feasible solution, the travel costs are irrelevant.

It is known (see Garey and Johnson [8]) that the 3-partitioning problem is $\mathcal{NP}$-hard in the strong sense, hence checking if a 1-PDTSP instance admits a feasible solution has the same computational complexity.

3   Transforming PDTSP into 1-PDTSP

The PDTSP involves the distribution of two commodities in a very special situation: one location (the depot) is the only source of a product and the only destination of the other product. This property allows us to solve PDTSP by using an algorithm for the 1-PDTSP. The transformation can be done as follows:

- duplicate the PDTSP depot into two dummy 1-PDTSP customers, one collecting all the PDTSP pickup-customer demands and the other providing all the PDTSP delivery-customer demands;

– impose that the vehicle goes directly from one dummy customer to the other with no load.

The second requirement is not necessary when the travel costs are Euclidean distances and $Q = \max\{\sum_{q_i > 0} q_i, -\sum_{q_i < 0} q_i\}$, as usually occurs in the test-bed instances of the PDTSP literature (see, e.g., Mosheiov [3] and Gendreau, Laporte and Vigo [9]).

4   Mathematical Model

This section presents an integer linear programming formulation for the 1-PDTSP. Let us start by introducing some notation. The depot will be denoted by 1 and each customer by i for $i = 2, \ldots, n$. To avoid trivial instances, we assume $n \geq 5$. For each pair of locations (i, j), the travel distance (or cost) c_{ij} of going from i to j is given. It is also given a demand q_i associated with each customer i, where $q_i < 0$ if i is a delivery customer and $q_i > 0$ if i is a pickup customer. In the case of a customer with zero demand, it is assumed that it is (say) a pickup customer that must be also visited by the vehicle. The capacity of the vehicle is represented by Q and is assumed to be a positive number. Let $V := \{1, 2, \ldots, n\}$ be the node set, A be the arc set between all nodes, and E the edge set between all nodes. For simplicity in notation, arc $a \in A$ with tail i and head j is also denoted by (i, j), and edge $e \in E$ with nodes i and j by $[i, j]$. For each subset $S \subset V$, let $\delta^+(S) := \{(i, j) \in A : i \in S, j \notin S\}$, $\delta^-(S) := \{(i, j) \in A : i \notin S, j \in S\}$ and $\delta(S) := \{[i, j] \in E : i \in S, j \notin S\}$.

Without loss of generality, the depot can be considered a customer by defining $q_1 := -\sum_{i=2}^{n} q_i$, i.e., a customer absorbing or providing the necessary amount of product to ensure product conservation. From now on we assume this simplification. Finally, let $K := \sum_{i \in V : q_i > 0} q_i = -\sum_{i \in V : q_i < 0} q_i$.

To provide a mathematical model to 1-PDTSP, for each arc $a \in A$, we introduce a 0-1 variable

$$x_a := \begin{cases} 1 & \text{if and only if } a \text{ is routed,} \\ 0 & \text{otherwise,} \end{cases}$$

and the continuous variable

$$f_a := \text{load of the vehicle going through arc } a.$$

Then the asymmetric 1-PDTSP can be formulated as:

$$\min \sum_{a \in A} c_a x_a$$

subject to

$$\sum_{a \in \delta^-(\{i\})} x_a = 1 \quad \text{for all } i \in V \tag{1}$$

$$\sum_{a \in \delta^+(\{i\})} x_a = 1 \quad \text{for all } i \in V \tag{2}$$

$$\sum_{a \in \delta^+(S)} x_a \geq 1 \quad \text{for all } S \subset V \tag{3}$$

$$x_a \in \{0,1\} \quad \text{for all } a \in A \tag{4}$$

$$\sum_{a \in \delta^+(\{i\})} f_a - \sum_{a \in \delta^-(\{i\})} f_a = q_i \quad \text{for all } i \in V \tag{5}$$

$$0 \leq f_a \leq Q x_a \quad \text{for all } a \in A. \tag{6}$$

From the model it is clear that the x_a variables in a 1-PDTSP solution represent a Hamiltonian cycle in a directed graph $G = (V, A)$, but not all Hamiltonian cycles in G define feasible 1-PDTSP solutions. Nevertheless, a trivial observation is that whenever a Hamiltonian cycle defined by a characteristic vector $[x'_a : a \in A]$ is a feasible 1-PDTSP solution (because there exists the appropriated loads $[f'_a : a \in A]$) then the cycle in the other direction, i.e.,

$$x''_{(i,j)} := \begin{cases} 1 & \text{if } x'_{(j,i)} = 1 \\ 0 & \text{otherwise} \end{cases} \quad \text{for all } (i,j) \in A,$$

is also a feasible 1-PDTSP solution. Indeed, vector $[x''_a : a \in A]$ admits the appropriated loads defined by:

$$f''_{(i,j)} := Q - f'_{(j,i)} \quad \text{for all } (i,j) \in A.$$

If $c_{ij} = c_{ji}$ for all $i, j \in V$ $(i \neq j)$, it is possible an smaller model by considering the new edge-decision variable

$$x_e := \begin{cases} 1 & \text{if and only if } e \text{ is routed} \\ 0 & \text{otherwise} \end{cases} \quad \text{for each } e \in E,$$

and a continuous non-negative variable g_a for each $a \in A$. Then the symmetric 1-PDTSP can be formulated as:

$$\min \sum_{e \in E} c_e x_e \tag{7}$$

subject to

$$\sum_{e \in \delta(\{i\})} x_e = 2 \quad \text{for all } i \in V \tag{8}$$

$$\sum_{e \in \delta(S)} x_e \geq 2 \quad \text{for all } S \subset V \tag{9}$$

$$x_e \in \{0,1\} \quad \text{for all } e \in E \tag{10}$$

$$\sum_{a \in \delta^+(\{i\})} g_a - \sum_{a \in \delta^-(\{i\})} g_a = q_i \quad \text{for all } i \in V \tag{11}$$

$$0 \leq g_{(i,j)} \leq \frac{Q}{2} x_{[i,j]} \quad \text{for all } (i,j) \in A. \tag{12}$$

Constraints (8) require that each customer must be visited once, and Constraints (9) require the 2-connectivity between customers. Constraints (11) and (12) guarantee the existence of a certificate $[g_a : a \in A]$ proving that a vector $[x_e : e \in E]$ defines a feasible 1-PDTSP cycle. In fact, if there exists a given Hamiltonian tour $[x'_e : e \in E]$ and a vector $[g'_a : a \in A]$ satisfying (8)–(12), then there also exists an oriented cycle $[x'_a : a \in A]$ and a vector $[f'_a : a \in A]$ satisfying the constraints in the original model (1)–(6), and thus guaranteeing a feasible 1-PDTSP solution. This can be done by simply considering any orientation of the tour defined by $[x'_e : e \in E]$ creating an oriented cycle represented by the characteristic vector $[x'_a : a \in A]$, and by defining

$$f'_{(i,j)} := g'_{(i,j)} + \left(\frac{Q}{2} - g'_{(j,i)} \right)$$

for each arc (i,j) in the oriented tour. Reciprocally, each feasible 1-PDTSP (i.e., each $[x'_a : a \in A]$ and $[f'_a : a \in A]$ satisfying (1)–(6)) corresponds to a feasible solution of model (7)–(12). Indeed, this solution is defined by:

$$x'_{[i,j]} := x'_{(i,j)} + x'_{(j,i)} \quad \text{for each } [i,j] \in E$$

and

$$g'_{(i,j)} := \begin{cases} f'_{(i,j)}/2 & \text{if } x'_{(i,j)} = 1, \\ (Q - f'_{(j,i)})/2 & \text{if } x'_{(j,i)} = 1, \\ 0 & \text{otherwise} \end{cases} \quad \text{for each } (i,j) \in A.$$

Therefore, the meaning of the continuous variable g_a in the symmetric 1-PDTSP model (7)–(12) is not properly the load of the vehicle going through arc a, as occurs in the asymmetric 1-PDTSP model, but it is a certificate that a Hamiltonian cycle $[x_e : e \in E]$ is a feasible 1-PDTSP solution.

Without the above observation, a quick overview could lead to the mistake of thinking that the upper limit of a load through arc (i,j) in constraints (12) must be $Qx_{[i,j]}$ instead of $Qx_{[i,j]}/2$. In other words, it would be a mistake to try to use variables f_a instead of g_a and to replace Constraints (11)–(12) by

$$\left\{ \begin{array}{ll} \displaystyle\sum_{a\in\delta^+(\{i\})} f_a - \sum_{a\in\delta^-(\{i\})} f_a = q_i & \text{for all } i \in V \\[2mm] 0 \le f_{(i,j)} \le Q x_{[i,j]} & \text{for all } (i,j) \in A. \end{array} \right\} \tag{13}$$

Indeed, let us consider an instance with a depot and two customers, one with demand $q_2 = +4$ and the other with demand $q_3 = -2$. If the vehicle capacity is (say) $Q = 2$ then the 1-PDTSP problem is not feasible, but the mathematical model (7)–(10) and (13) has the integer solution $x_{[1,2]} = x_{[2,3]} = x_{[1,3]} = 1$ with $f_{(1,2)} = 0, f_{(2,3)} = 2, f_{(3,1)} = 1$ and $f_{(1,3)} = 1, f_{(3,2)} = 0, f_{(2,1)} = 2$. Therefore, model (7)–(10) and (13) is not valid.

By Benders' decomposition (see Benders [7]) it is possible to project out the continuous variables g_a in model (7)–(12), obtaining a pure 0-1 ILP model on the decision variables. Indeed, a given Hamiltonian cycle $[x_e : e \in E]$ defines a feasible 1-PDTSP solution if there exists a vector $[g_a : a \in A]$ satisfying (11) and (12). According to Farkas' Lemma the polytope described by (11) and (12) for a fixed vector $[x_e : e \in E]$ is feasible if, and only if, all extreme directions $[\alpha_i : i \in V , \ \beta_a : a \in A]$ of the polyhedral cone

$$\left\{ \begin{array}{ll} \alpha_i - \alpha_j \le \beta_{(i,j)} & \text{for all } (i,j) \in A \\[1mm] \beta_a \ge 0 & \text{for all } a \in A \end{array} \right\} \tag{14}$$

satisfy

$$\sum_{i\in V} \alpha_i q_i - \sum_{(i,j)\in A} \beta_{(i,j)} \frac{Q}{2} x_{[i,j]} \le 0.$$

Clearly, the linearity space of (14) is a 1-dimensional space generated by the vector defined by $\tilde{\alpha}_i = 1$ for all $i \in V$ and $\tilde{\beta}_a = 0$ for all $a \in A$. Therefore, it is possible to assume $\alpha_i \ge 0$ for all $i \in V$ in (14) to simplify the characterization of the extreme rays. In fact, this assumption also follows immediately from the fact that equalities "=" in the linear system (11) can be replaced by inequalities "$\ge$" without adding new solutions. Therefore, let us characterize the extreme rays of the cone:

$$\left\{ \begin{array}{ll} \alpha_i - \alpha_j \le \beta_{(i,j)} & \text{for all } (i,j) \in A \\[1mm] \alpha_i \ge 0 & \text{for all } i \in V \\[1mm] \beta_a \ge 0 & \text{for all } a \in A. \end{array} \right\} \tag{15}$$

Theorem 1. *Except for multiplication by positive scalars, all the extreme directions of the polyhedral cone defined by (15) are:*

(i) for each $a' \in A$, the vector (α, β) defined by $\alpha_i = 0$ for all $i \in V$, $\beta_a = 0$ for all $a \in A \setminus \{a'\}$ and $\beta_{a'} = 1$;

(ii) for each $S \subset V$, the vector (α, β) defined by $\alpha_i = 1$ for all $i \in S$, $\alpha_i = 0$ for all $i \in V \setminus S$, $\beta_a = 1$ for all $a \in \delta^+(S)$ and $\beta_a = 0$ for all $a \in A \setminus \delta^+(S)$.

Proof: It is evident that the two families of vectors are directions of (15). First, we will show that they are also extreme vectors. Second, we will show that they are the unique ones in (15).

Clearly vectors of family (i) are extreme. Let us now consider a vector (α, β) of family (ii), and two other vectors (α', β') and (α'', β'') satisfying (15) such that

$$(\alpha, \beta) = \frac{1}{2}(\alpha', \beta') + \frac{1}{2}(\alpha'', \beta'').$$

By definition, $\alpha_i = 0$ for all $i \notin S$ and $\beta_a = 0$ for all $a \notin \delta^+(S)$. Because of the non-negativity assumption on all the components, it follows $\alpha'_i = 0 = \alpha''_i$ for all $i \notin S$ and $\beta'_a = 0 = \beta''_a$ for all $a \notin \delta^+(S)$. Considering $a = (i,j)$ with $i, j \in S$, the last result implies $\alpha'_i = \alpha'_j$ and $\alpha''_i = \alpha''_j$. Moreover, for all $(i,j) \in \delta^+(S)$, $\alpha_i = 1$ and $\beta_{(i,j)} = 1$, thus $\alpha'_i + \alpha''_i = 2 = \beta'_{(i,j)} + \beta''_{(i,j)}$ whenever $(i,j) \in \delta^+(S)$. Since $\alpha'_i \leq \beta'_{(i,j)}$ and $\alpha''_i \leq \beta''_{(i,j)}$ then $\alpha'_i = \beta'_{(i,j)}$ and $\alpha''_i = \beta''_{(i,j)}$ whenever $(i,j) \in \delta^+(S)$. In conclusion, (α', β') and (α'', β'') coincide with (α, β), thus (α, β) is an extreme direction of the cone described by (15).

Let us now prove that the two families of vectors are *all* the extreme directions of the cone defined by (15). To do that, let us consider any vector (α', β') satisfying (15) and let us prove that it is a cone combination of vectors involving at least one from the two families. The aim is obvious when $\alpha' = 0$ by considering the vectors of family (i), so let us assume that α' has a positive component. Set

$$S' := \{i \in V : \alpha'_i > 0\} \qquad \text{and} \qquad \lambda' := \min\{\alpha'_i : i \in S'\}.$$

Let us consider (α'', β'') the vector in family (ii) defined by $S := S'$, and $(\alpha''', \beta''') := (\alpha', \beta') - \lambda'(\alpha'', \beta'')$. The proof concludes since $\lambda' > 0$ and (α''', β''') satisfies (15).   □

Therefore, a given Hamiltonian cycle $[x_e : e \in E]$ defines a feasible 1-PDTSP solution if and only if

$$\sum_{e \in \delta(S)} x_e \geq \frac{2}{Q} \sum_{i \in S} q_i$$

for all $S \subset V$ (case $S = V$ is unnecessary). These inequalities are known in the CVRP literature as *capacity constraints* (see, e.g., Toth and Vigo [19]). Since $\delta(S) = \delta(V \setminus S)$ and $\sum_{i \in S} q_i = \sum_{i \notin S}(-q_i)$, the above inequalities are equivalent to

$$\sum_{e \in \delta(S)} x_e \geq \frac{2}{Q} \sum_{i \in S}(-q_i)$$

for all $S \subset V$.

An immediate consequence is that 1-PDTSP can be formulated as the classical TSP model (7)–(10) plus the following *Benders' cuts*:

$$\sum_{e \in \delta(S)} x_e \geq \frac{2}{Q} \left| \sum_{i \in S} q_i \right| \qquad \text{for all } S \subseteq V : |S| \leq |V|/2. \tag{16}$$

Even if there is an exponential number of linear inequalities in (16), today's state-of-the-art of cutting-plane approaches allows us to manage all of them in a very effective way. To be more precise, as Section 5.1 discusses in detail, Constraints (16) can be efficiently incorporated by finding (if any) a feasible solution for the linear system (11)–(12) with $[x_e : e \in E]$ as parameters. Therefore, any cutting-plane approach for solving the TSP can be adapted to solve the 1-PDTSP by also considering Constraints (16). This means that it could be possible to insert the new inequalities in a software like CONCORDE (see Applegate, Bixby, Chvatal and Cook [4]) and to obtain an effective program to solve instances of 1-PDTSP. Unfortunately, the source code of this particular TSP software is very complex to modify and we did not succeed in the adaptation. That was a motivation to develop the "ad hoc" implementation described in the next section.

5   Algorithm for 1-PDTSP

In this section an enumerative algorithm is proposed for the exact solution of the problem. The algorithm follows a branch-and-bound scheme, in which lower bounds are computed by solving an LP relaxation of the problem. The relaxation is iteratively tightened by adding valid inequalities to the current LP, according to the so-called *cutting plane* approach. The overall method is commonly known as a *branch-and-cut* algorithm; refer to Padberg and Rinaldi [17] and Jünger, Reinelt and Rinaldi [12] for a thorough description of the technique. Some important implementation issues are next described.

5.1   Separating Benders' Cuts

Due to the large number of inequalities in (16), not all of them can be considered in an LP relaxation of the problem. Useful constraints must be identified dynamically, and this is typically called *the separation problem of (16)*: "given a (possibly fractional) solution $[x_e : e \in E]$ of an LP relaxation, is there a violated cut in (16)? If yes, provide with (at least) one".

As it was mentioned in Section 4, this question can be answered by checking the feasibility of the polytope described by (11)–(12). Indeed, in the Benders' decomposition terminology, the problem of finding (if any) a vector $[g_a : a \in A]$ satisfying (11)–(12) for a given $[x_e : e \in E]$ is called *subproblem*. An easy way to solve the subproblem is to solve a linear program (LP) by considering (11)–(12) with a dummy objective function. If it is not feasible then its dual program is unbounded and the unbounded extreme direction defines a violated Benders' cut; otherwise, all constraints in (16) are satisfied. Therefore the separation problem of (16) can be solved in polynomial time by using, for example, an implementation of the ellipsoid method for Linear Programming (see [13]). Nevertheless, in practice the efficiency of the overall algorithm strongly depends on this phase. A better way of solving the separation problem for Constraints (16) follows the idea presented by Harche and Rinaldi (see [5]) and is next addressed.

Let $[x_e^* : e \in E]$ be a given solution of a linear relaxation of model (7)–(10) and (16). In order to check whether there is a violated Benders' cut, let us write the constraints in a different form. For each $S \subset V$

$$\sum_{e \in \delta(S)} x_e^* \geq \sum_{i \in S} \frac{+2q_i}{Q}$$

is algebraically equivalent to:

$$\sum_{e \in \delta(S)} x_e^* + \sum_{i \in V \setminus S : q_i > 0} \frac{2q_i}{Q} + \sum_{i \in S : q_i < 0} \frac{-2q_i}{Q} \geq \sum_{i \in V : q_i > 0} \frac{2q_i}{Q}.$$

The right-hand side of the inequality is the positive constant $2K/Q$, and the coefficients in the left-hand side are also non-negative. Therefore, this result produces an algorithm to solve the separation problem of the Benders' cuts based on solving a max-flow problem on the capacitated undirected graph $G^* = (V^*, E^*)$ defined as follows.

Consider two dummy nodes $n + 1$ and $n + 2$, and let $V^* := V \cup \{n + 1, n + 2\}$. The edge set E^* contains the edges $e \in E$ such that $x_e^* > 0$ in the given solution with capacity x_e^*, the edge $[i, n + 1]$ for each pickup customer i with capacity $2q_i/Q$, and the edge $[i, n+2]$ for each delivery customer i with capacity $-2q_i/Q$.

Finding a most-violated Benders' inequality in (16) calls for the minimum-capacity cut $(S^*, V^* \setminus S^*)$ with $n + 1 \in S^*$ and $n + 2 \in V^* \setminus S^*$ in the capacitated undirected graph G^*. This can be done in $O(n^3)$ time, as it amounts to finding the maximum flow from $n + 1$ to $n + 2$ (see Ahuja, Magnanti and Orlin [1]). If the maximum flow value is no less than $2K/Q$ then all the inequalities (16) are satisfied; otherwise the capacity of the minimum cut separating $n + 1$ and $n + 2$ is strictly less than $2K/Q$ and a most-violated inequality (16) has been detected among all. The subset S defining a most-violated Benders' inequality is either $S^* \setminus \{n + 1\}$ or $V^* \setminus (S^* \cup \{n + 2\})$.

5.2   Strengthening the LP-Relaxation

Even if a solution $[x_e^* : e \in E]$ satisfies all constraints of the LP relaxation of model (7)–(10) and (16), its objective value can be even farther from the objective value of an optimal 1-PDTSP solution. Therefore, it is always important to provide ideas to strengthen the LP relaxation and ensure better lower bounds. Some ideas are next addressed.

A first improvement of an LP relaxation arises by rounding up to the next even number the right-hand side of constraints (16), i.e., to consider the following *rounded Benders' cuts*:

$$\sum_{e \in \delta(S)} x_e \geq 2r_1(S) \qquad \text{for all } S \subseteq V, \tag{17}$$

where

$$r_1(S) := 2 \max \left\{ 1, \left\lceil \frac{|\sum_{i \in S} q_i|}{Q} \right\rceil \right\}.$$

This lifting of the right-hand-side in Constraints (16) is possible because the left-hand-side represents the number of times the vehicle goes through $\delta(S)$, and this is always a positive even integer. Unfortunately, the polynomial procedures described in Section 5.1 to solve the separation problem for (16) cannot be easily adapted to find a most-violated rounded Benders' cut. Nevertheless, it is very useful to insert in practice a Benders' cut in the rounded form whenever a constraint (9) or (16) is separated using the polynomial procedures.

A further constraint improvement of (9) and (16) in the 1-PDTSP formulation can be obtained by defining $r_2(S)$ as the smallest number of times the vehicle with capacity Q must go inside S to meet the demand q_i of the customers in S. The new valid inequalities are the following:

$$\sum_{e \in \delta(S)} x_e \geq 2r_2(S) \qquad \text{for all } S \subseteq V. \tag{18}$$

Notice that $r_2(S)$ is not the solution of a Bin Packing Problem since q_i is allowed to be negative. Observe that $r_1(S) \leq r_2(S)$ and inequality can hold strictly as in the following example. Let $S = \{1, 2, 3, 4\}$ a set of four customers with demands $q_1 = q_2 = +3$ and $q_3 = q_4 = -2$. If $Q = 3$ then $r_1(S) = 1$ and $r_2(S) = 2$. The computation of $r_2(S)$, even for a fixed subset S, is $\mathcal{NP}$-hard problem in the strong sense (see Section 2). Therefore, we do not consider constraints (18) in our algorithm.

Even if the computation of $r_1(S)$ is trivial for a given $S \subset V$, it is very unlikely to find a polynomial algorithm to solve the separation problem of (17) (similar constraints were proven to have an $\mathcal{NP}$-complete separation problem by Harche and Rinaldi for the Capacitated Vehicle Routing Problem; see [5]). Therefore, we have implemented some simple heuristic approaches to separate (17) in a similar way as the one described in [5] for the CVRP:

- The first heuristic procedure looks for the most violated constraint (16) by using the exact procedure described in Section 5.1. If S^* is the output of this procedure, constraints (17) is checked for violation.
- The second procedure compares the current best feasible solution and the solution of the current LP relaxation. By removing edges in the best feasible cycle associated with variables close to value 1 in the current LP solution, the resulting connected components are considered as potential node sets for defining violated rounded Benders' constraints.
- Whenever a subset S' defining a violated constraint is identified, a third procedure checks for violation the inequality (17) associated to subset $S'' :=$

$S' \cup \{v\}$ for each $v \notin S'$ and $S'' := S' \setminus \{v\}$ for each $v \in S'$. Notice that when $r_1(S'') > r_1(S')$ then constraint (17) defined by $S = S''$ dominates the one defined by $S = S'$.

These heuristic approaches will be denoted as Sep1, Sep2 and Sep3, respectively.

Another strengthening arises by considering all valid inequalities known for the TSP. Indeed, as mentioned before, the 1-PDTSP is a TSP plus additional constraints, so all constraints (e.g., 2-matching inequalities) can be used to improve the lower bound from LP relaxations of 1-PDTSP. See Naddef [16] for the more common facet-defining inequalities of the TSP polytope.

Generalizing this idea, it is also possible to use polyhedral results known for the CVRP. Indeed, constraints (16), (17) and (18) are extensions of the so-called *capacity constraints* in the CVRP literature (see, e.g., Toth and Vigo [19]). In the same way, other inequalitites can be adapted for the 1-PDTSP. This is the case of the *multistar constraints* (see, e.g., Gouveia [14]) that for the 1-PDTSP are:

$$\sum_{e \in \delta(S)} x_e \geq \frac{2}{Q} \left| \sum_{i \in S} \left(q_i + \sum_{j \in V \setminus S} q_j x_{[i,j]} \right) \right| \tag{19}$$

for all $S \subset V$. The validity follows from the observation that each time the vehicle visits S and uses edge $[i,j]$ with $i \in S$ and $j \notin S$, the vehicle must have free capacity for visiting the nodes in S and also for j. Different from the situation in the CVRP, multistar inequalities (19) does not necessarily dominate the capacity inequalities (16). As for the separation, it is clear that the procedure described in Section 5.1 can be easily adapted to solve in polynomial time the separation of (19) when $q_j \leq Q/2$ for all $j \in V$.

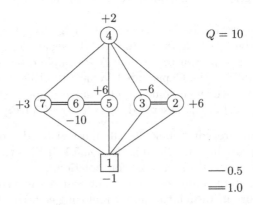

Fig. 1. Fractional solution of model (7)–(10) and (17)

A final important improvement is based on the existence of incompatibilities between some edges in the graph G. Figure 1 shows an example of a typical fractional vector $[x_e : e \in E]$ satisfying all linear constraints in (7)–(10) and (17). Edges in single lines represent variables with value 0.5, edges in double lines represent variables with value 1, and no-drawn edges represent variables with value 0. This fractional solution is a convex combination of two Hamiltonian cycles, characterized by the node sequences $(1, 2, 3, 4, 5, 6, 7, 1)$ and $(1, 3, 2, 4, 7, 6, 5, 1)$, where the first is a feasible 1-PDTSP but the second is not. Clearly the vehicle capacity requirement forces some variables to be fixed at zero. This is the case of the variable associated to edge $[1, 6]$. Moreover, Q also produces incompatibilities between pairs of edge variables. In the example, each pair of edges in $\{[2, 4], [4, 5], [4, 7]\}$ are incompatible. Indeed, the vehicle cannot route e.g. $[2, 4]$ and $[4, 5]$ consecutively since $8 + 8 - 2 > Q$. This can be mathematically written as

$$x_{[2,4]} + x_{[4,5]} \leq 1 \quad , \quad x_{[2,4]} + x_{[4,7]} \leq 1 \quad , \quad x_{[4,5]} + x_{[4,7]} \leq 1.$$

None of the above inequalities is violated by the fractional solution in Figure 1, but there is a stronger way of imposing the 3-pair incompatibility:

$$x_{[2,4]} + x_{[4,5]} + x_{[4,7]} \leq 1,$$

which is a violated cut by the fractional solution.

Therefore, in 1-PDTSP with low capacity Q, there is an underlying Set-Packing structure and all known valid inequalities can be used to strengthening the LP relaxation of the 1-PDTSP. The interested reader is referred to Balas and Padberg [6] for a survey on the *Set-Packing Problem*. Basic constraints of the Set-Packing polytope are the so-called *clique inequalities* defined in the problem as follows. Consider the so-called *conflict graph* $G^c = (V^c, E^c)$ where there is a node in V^c for each edge-decision variable, and an edge in E^c connecting two nodes when the corresponding variables can not be consecutively routed by the vehicle. Then each subset $W \subseteq V^c$ inducing a complete subgraph of G^c defines the following clique inequality:

$$\sum_{e \in W} x_e \leq 1. \tag{20}$$

Only inequalities associated to maximal-inclusion cliques are computationally useful (and also facet-defining of the Set-Packing polytope). Since finding a maximal clique in a general graph is an $\mathcal{NP}$-hard problem, we heuristically solve the separation problem of (20). Our procedure consists of a simple exhaustive search of stars of the support graph associated G^* to the solution. In particular, given a fractional solution $[x_e^* : e \in E]$ of an LP relaxation of 1-PDTSP, and for each customer i, we enumerate all three edges in $\delta(i)$ in G^* defining a violated clique inequality.

5.3   Heuristic Algorithms

To speed up the branch-and-bound algorithm it is very important not only to have good lower bounds but also good feasible solutions. In order to achieve this second aim, we have developed two main heuristic algorithms: an *initial heuristic* to be executed at the beginning of the enumerative algorithm, and a *primal heuristic* to be executed after each node of the branch-and-bound tree using the solution of the current LP relaxation.

A number of known tour construction and tour improvement heuristic algorithms for TSP (see, e.g., Golden and Stewart [10]) can be easily adapted to 1-PDTSP, even if in this problem we can not always ensure that a tour construction procedure ends with a feasible 1-PDTSP cycle. Indeed, we have implemented nearest insertion, farthest insertion, cheapest insertion, two-optimality and three-optimality TSP-like procedures, trying to guarantee feasibility and low-cost in the final solutions. See [11] for more details on each procedure.

5.4   Branching

When the solution $[x_e^* : e \in E]$ from the LP-relaxation had non-integer values, we considered the branching on variables, the standard approach for branch-and-cut. It consists of selecting a fractional edge-decision variable x_e for generating two descendant nodes by fixing x_e to either 0 or 1. In our implementation we chose the variable with value x_e^* as close as possible to 0.5 (ties are broken by choosing the edge having maximum cost). We performed also experiments with a branching scheme based on subsets (selecting a subset S previously generated within a separation procedure and such that $\sum_{e \in \delta(S)} x_e^*$ was as close as possible to an odd number) but we obtained worse results in our benchmark instances. See [11] for more details.

6   Preliminary Computational Results

The enumerative algorithm described in Section 5 has been implemented in ANSI C, and it ran on a personal computer AMD 1333 Mhz. As to the LP solver the package CPLEX 7.0 was used. To test the performance of our approach both on the 1-PDTSP and on the PDTSP, we have considered a classical test-bed PDTSP instance introduced in Mosheiov [15]. It consists of the depot, 12 pickup customers and 12 delivery customers. The capacity of the vehicle in the PDTSP instance is given by $Q := \max\{\sum_{i \in V : q_i > 0} q_i, -\sum_{i \in V : q_i < 0} q_i\} = 45$ and the cost c_{ij} by the Euclidean distance between points i and j. By solving the TSP on this instance to optimality, we got the Hamiltonian cycle illustrated in [15] but with a different objective value (we obtained 4431 while Mosheiov got 4434). We tried different roundings of the Euclidean distances but we did not succeed in obtaining the same optimal TSP value. In our final implementation we computed the costs using the same code instruction from CONCORDE [4].

We also generated 1-PDTSP instances by using the same demands and distances of the Mosheiov instance and reducing the capacity of the vehicle. In particular we noticed that when $Q \geq 16$ an optimal solution of the 1-PDTSP instance is an optimal TSP solution using the costs c_{ij}, while there are a customer with demand 7. Therefore, we generated ten instances, one for each $Q \in \{7, 8, \ldots, 15, 16\}$.

Table 1 summarizes the results of our experiments. For each instance, the meaning of a column is as follows:

Sep1 : shows the percentage ratio of the lower bound over the optimal objective value, using the separation procedure Sep1 described in Section 5.1;

Sep2 : shows the percentage ratio of the lower bound when the heuristic procedure Sep2 described in Section 5.2 is applied;

Sep3 : shows the percentage ratio of the lower bound when the heuristic separation procedure Sep3 described in Section 5.2 is also applied;

r-LB: shows the percentage ratio after considering some clique inequalities according with the heuristic separation procedure described in Section 5.2;

UB_0: is the percentage ratio of the initial heuristic described in Section 5.3;

Optimum: is the optimal objective function value (and denominator in ratios);

Heu: is the time in seconds of the AMD 1333 Mhz. to perform the heuristic procedures;

Root: is the time in seconds of the AMD 1333 Mhz. to perform the root node of the branch-and-bound tree excluding the heuristic procedures;

Time: is the overall time of the algorithm in seconds of the AMD 1333 Mhz;

Deep: is the maximum level explored in the branch-and-bound tree;

Nodes: is the number of explored branch-and-bound nodes.

Table 1. Ten 1-PDTSP instances and the PDTSP instance from data in [15]

n	Q	Sep1	Sep2	Sep3	r-LB	UB_0	Optimum	Heu	Root	Time	Deep	Nodes
25	7	88.18	93.43	94.16	95.36	100.00	5734	0.17	0.16	3.46	10	155
25	8	90.43	97.49	97.51	98.00	100.00	5341	0.17	0.11	0.82	7	35
25	9	92.30	96.92	98.85	98.85	100.00	5038	0.12	0.11	0.33	2	5
25	10	88.55	99.40	99.44	99.44	100.00	4979	0.05	0.11	0.16	2	5
25	11	91.59	97.76	99.88	99.88	100.00	4814	0.06	0.11	0.17	1	3
25	12	91.59	96.41	97.76	97.76	100.00	4814	0.06	0.05	0.17	2	5
25	13	95.29	98.88	99.61	99.61	100.00	4627	0.05	0.05	0.16	1	3
25	14	98.50	100.00	100.00	100.00	100.00	4476	0.06	0.05	0.11	0	1
25	15	98.50	100.00	100.00	100.00	100.00	4476	0.05	0.06	0.11	0	1
25	16	99.50	99.50	99.50	99.50	100.00	4431	0.06	0.05	0.11	3	7
25	45	100.00	100.00	100.00	100.00	100.00	4467	0.06	0.06	0.17	0	1

Table 1 shows that, on the Mosheiov data, the complexity of the classical PDTSP is similar to the classical TSP, while the 1-PDTSP turns out to be harder when Q decreases. In fact, we found 26 and 10 violated cliques inequalities when $Q = 7$ and $Q = 8$, respectively, while none was found when $Q \geq 9$. Without the clique inequality separation the number of explored branch-and-bound nodes to solve the instance with $Q = 7$ increases from 117 to 279 and the computational time from 106.5 to 244.5 seconds. The initial heuristic approach consumed 0.2 seconds on each instance and the quality of the provided solution was quite good. As a final observation, the heuristic PDTSP solution in Mosheiov [15] has an objective value of 4635 (using our cost matrix, and 4634 in [15]) while an optimal PDTSP solution has a value of 4467. See [11] for other computational experiments on the randomly generated instances described in [15] and [9], where the behaviors of the overall algorithm is similar.

References

1. R.K. Ahuja, T.L. Magnanti, J.B. Orlin, "Network Flows", in G.L. Nemhauser, A.H.G. Rinnooy Kan, M.J. Todd (Editors), "Optimization" I, North-Holland, 1989.
2. S. Anily, R. Hassin, "The Swapping Problem", *Networks* 22 (1992) 419–433.
3. S. Anily, G. Mosheiov, "The traveling salesman problem with delivery and backhauls", *Operations Research Letters* 16 (1994) 11–18.
4. D. Applegate, R. Bixby, V. Chvatal, W. Cook, "Concorde: a code for solving Traveling Salesman Problem", http://www.math.princeton.edu/tsp/concorde.html, 1999.
5. P. Augerat, J.M. Belenguer, E. Benavent, A. Corberán, D. Naddef, G. Rinaldi, "Computational Results with a Branch and Cut Code for the Capacitated Vehicle Routing Problem", Research Report 949-M, 1995, Universite Joseph Fourier, Grenoble, France.
6. E. Balas, M.W. Padberg, "Set Partitioning: A survey", *SIAM Review* 18 (1976) 710–760.
7. J.F. Benders, "Partitioning Procedures for Solving Mixed Variables Programming Problems", *Numerische Mathematik* 4 (1962) 238–252.
8. M.R. Garey, D.S. Johnson, "Computers and Intractability: A Guide to the Theory of $\mathcal{NP}$-Completeness", *W.H. Freeman and Co.*, 1979.
9. M. Gendreau, G. Laporte, D. Vigo, "Heuristics for the traveling salesman problem with pickup and delivery", *Computers & Operations Research* 26 (1999) 699–714.
10. B.L. Golden, W.R. Stewart, "Empirical analysis of heuristics", in E.L. Lawler, J.K. Lenstra, A.H.G. Rinnooy Kan, D.B. Shmoys (Editors), *The Traveling Salesman Problem: A Guided Tour of Combinatorial Optimization*, Wiley, 1985.
11. H. Hernández, J.J. Salazar, "A Branch-and-Cut Algorithm for the Pickup-and-Delivery Travelling Salesman Problem", technical report, University of La Laguna, 2001. Available on http://webpages.ull.es/users/jjsalaza/pdtsp.htm.
12. M. Jünger, G. Reinelt, G. Rinaldi, "The Travelling Salesman Problem", in M.O. Ball, T.L. Magnanti, C.L. Monma, G.L. Nemhauser (Editors), *Handbook in Operations Research and Management Science: Network Models*, Elsevier, 1995.
13. L.G. Khachian, "A polynomial Algorithm in Linear Programming", *Soviet Mathematics Doklady* 20 (1979) 191–194.
14. L. Gouveia, "A result on projection for the the Vehicle Routing Problem", *European Journal of Operational Research* 83 (1997) 610–624.
15. G. Mosheiov, "The Travelling Salesman Problem with pick-up and delivery", *European Journal of Operational Research* 79 (1994) 299–310.
16. D. Naddef, "Polyhedral theory and Branch-and-Cut algorithms for Symmetric TSP", Chapter 4 in G. Gutin, A. Punnen (Editors), *The Traveling Salesman Problem and its Variants*, Kluwer, 2001.
17. M.W. Padberg, G. Rinaldi, "A branch-and-cut algorithm for the resolution of large-scale symmetric traveling salesman problems", *SIAM Review* 33 (1991) 60–100.
18. M.W.P. Savelsbergh, M. Sol, "The General Pickup and Delivery Problem", *Transportation Science* 29 (1995) 17–29.
19. P. Toth, D. Vigo (Editors), "The Vehicle Routing Problem", *SIAM Discrete Mathematics and its Applications*, 2001.

Reconstructing a Simple Polytope from Its Graph

Volker Kaibel[*]

TU Berlin, MA 6–2
Straße des 17. Juni 136
10623 Berlin, Germany
kaibel@math.tu-berlin.de
http://www.math.tu-berlin.de/~kaibel

Abstract. Blind and Mani [2] proved that the entire combinatorial structure (the vertex-facet incidences) of a simple convex polytope is determined by its abstract graph. Their proof is not constructive. Kalai [15] found a short, elegant, and algorithmic proof of that result. However, his algorithm has always exponential running time. We show that the problem to reconstruct the vertex-facet incidences of a simple polytope P from its graph can be formulated as a combinatorial optimization problem that is strongly dual to the problem of finding an abstract objective function on P (i.e., a shelling order of the facets of the dual polytope of P). Thereby, we derive polynomial certificates for both the vertex-facet incidences as well as for the abstract objective functions in terms of the graph of P. The paper is a variation on joint work with Michael Joswig and Friederike Körner [12].

1 Introduction

The *face lattice* L_P of a (convex) polytope P is any lattice that is isomorphic to the lattice formed by the set of all faces of P (including $\varnothing$ and P itself), ordered by inclusion. It is well-known to be determined by the *vertex-facet incidences* of P, i.e., by any graph that is isomorphic to the bipartite graph whose nodes are the vertices and the facets of P, where the edges are defined by the pairs $\{v, f\}$ of vertices v and facets f with $v \in f$. In lattice theoretic terms, L_P is a ranked, atomic, and coatomic lattice, and thus, the sub-poset formed by its atoms and coatoms already determines the whole lattice. Actually, one can compute L_P from the vertex-facet incidences of P in $\mathcal{O}(\eta \cdot \alpha \cdot \lambda)$ time, where η is the minimum of the number of vertices and the number of facets, α is the number of vertex-facet incidences, and λ is the total number of faces of P [14].

The *graph* $G_P = (V_P, E_P)$ of a polytope P is any graph that is isomorphic to the graph whose nodes are the vertices of P, where two nodes are adjacent if and only if the convex hull of the corresponding two vertices is a one-dimensional face of P. Phrased differently, G_P is the graph defined on the rank one elements of L_P,

---

[*] Supported by the Deutsche Forschungsgemeinschaft, FOR 413/1–1 (Zi 475/3–1).

M. Jünger et al. (Eds.): Combinatorial Optimization (Edmonds Festschrift), LNCS 2570, pp. 105–118, 2003.
© Springer-Verlag Berlin Heidelberg 2003

where two rank one elements are adjacent if and only if they are below a common rank two element. While the vertex-facet incidences completely determine the face-lattice of any polytope, the graph of a polytope in general does not encode the entire combinatorial structure. This can be seen, e.g., from the examples of the cut polytope associated with the complete graph on n nodes and the $\binom{n}{2}$-dimensional cyclic polytope with 2^{n-1} vertices, which both have complete graphs. Another example is the four-dimensional polytope shown in Fig. 1 whose graph is isomorphic to the graph of the five-dimensional cube.

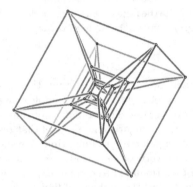

Fig. 1. A Schlegel-diagram (projection onto one facet) of a four-dimensional polytope with the graph of a five-dimensional cube, found by Joswig & Ziegler [13].

Actually, in all dimensions less than four such ambiguities cannot occur. For one- or two dimensional polytopes this is obvious, and for three-dimensional polytopes it follows from Whitney's theorem [18] saying that every 3-connected planar graph has a unique (up to reflection) plane embedding.

A d-dimensional polytope P is *simple* if every vertex of P is contained in precisely d facets, which is equivalent to G_P being d-regular (the polytope is non-degenerate in terms of Linear Programming). Every face of a simple polytope is simple as well. None of the examples showing that it is, in general, impossible to reconstruct the face lattice of a polytope from its graph, is simple.

In fact, Blind and Mani [2] proved in 1987 that the face lattice of a simple polytope is determined by its graph. Their proof (which we sketch in Sect. 2) is not constructive and crucially relies on the topological concept of homology. In 1988, Kalai [15] found a short and elegant proof (reviewed in Sect. 3) that does only use elementary geometric and combinatorial reasoning with the main advantage of being algorithmic. However, the running time of the method that can be devised from it is exponential in the size of the graph.

Perles conjectured in the 1970's (see [15]) that for a d-dimensional simple polytope P every subset $F \subset V_P$ that induces a $(d-1)$-regular, connected, and non-separating subgraph of G_P corresponds to the vertex set of a facet of P. A proof of this conjecture would have lead immediately to a polynomial time algorithm that, given the graph $G_P = (V_P, E_P)$ of a simple polytope P, decides for a set of subsets of V_P if it corresponds to the set of vertex sets of facets of P. However, Haase and Ziegler [10] recently disproved Perles' conjecture. They found a four-dimensional simple polytope whose graph has a 3-regular, non-separating, and even 3-connected induced subgraph that does not correspond to any facet.

Refining ideas from Kalai's proof (Sect. 4), we show that the problem of reconstructing the vertex-facet incidences of a simple polytope P from its graph G_P can be formulated as a combinatorial optimization problem that has a well-stated strongly dual problem (Sect. 5). The optimal solutions to this dual problem are certain orientations of G_P (induced by "abstract objective functions") that are important also in different contexts. In particular, we provide short certificates for both the vertex-facet incidences of a simple polytope and for the abstract objective functions in terms of G_P. We conclude with some remarks on the complexity status of the problem to decide whether a claimed solution to the reconstruction problem indeed are the vertex-facet incidences of the respective polytope in Sect. 6.

The material presented here has evolved from joint work with Michael Joswig and Friederike Körner [12]. The basic ideas and results are the same in both papers. However, the concept of a "facoidal system of walks" is newly introduced here. It differs from the corresponding notion of a "2-system" (introduced in [12]) with the main effect that one knows how to compute efficiently some facoidal system of walks from the graph G_P of a simple polytope P (see Proposition 3), while it is unclear how to find a 2-system from G_P efficiently. Furthermore, the proof of Theorem 3 we give here is different from the corresponding proof in [12]. Finally, the complexity theoretic statement of Corollary 5 does not appear in [12].

For all notions from the theory of polytopes that we use without (sufficient) explanations, we refer to Ziegler's book [19]. We use the terms *d-polytope* and *k-face* for d-dimensional polytopes and k-dimensional faces, respectively. Often we will identify a face F of a simple polytope P with the subset of nodes of G_P that corresponds to the vertex set of F. Whenever we talk about "polynomial time" or "efficient" this refers to the size of the graph G_P of the respective (simple) polytope P.

2   The Theorem of Blind and Mani

Blind and Mani [2] proved their theorem in the dual setting, i.e., for simplicial rather than for simple polytopes. Nevertheless, we sketch parts of their proof in terms of simple polytopes here. The starting point is the observation that, while a priori it is by no means clear if the graph $G_P = (V_P, E_P)$ of a simple polytope P determines the face lattice of P, it is easy to see that the 2-faces of P (as subsets of V_P) carry the entire information on the combinatorial structure of P.

Let P be a simple polytope. For a node $v \in V_P$ of G_P denote by $\Gamma(v) \subset V_P$ the subset of nodes that are adjacent to v (the *neighbors* of v). For any k-element subset $S \subset \Gamma(v)$ there is a k-face of P that contains the vertices that correspond to $S \cup \{v\}$ (and no vertex that corresponds to a node in $\Gamma(v) \setminus S$). We call the subset $F(S \cup \{v\}) \subseteq V_P$ of nodes corresponding to the vertices of that face the *k-face spanned by* $S \cup \{v\}$.

For an edge $e = \{v, w\} \in E_P$ let $\Psi_{(v,w)} : \Gamma(v) \setminus \{w\} \longrightarrow \Gamma(w) \setminus \{v\}$ be the map that assigns to each subset $S \subset \Gamma(v) \setminus \{w\}$ the subset $T \subset \Gamma(w) \setminus \{v\}$ with $F(S \cup \{v, w\}) = F(T \cup \{w, v\})$. The maps $\Psi_{(v,w)}$ are cardinality preserving bijections, where $\Psi_{(w,v)}$ is the inverse of $\Psi_{(v,w)}$.

Proposition 1. *Let P be a simple polytope. For each $e = \{v, w\} \in E_P$ and $S \subseteq \Gamma(v) \setminus \{w\}$ we have*

$$\Psi_{(v,w)}(S) = \overline{\Psi_{(v,w)}(\overline{S})}$$

(where $\overline{U}$ is the respective complement of the set U).

Proof. This follows from the fact that we have $\Psi_{(v,w)}(S_1 \cap S_2) = \Psi_{(v,w)}(S_1) \cap \Psi_{(v,w)}(S_2)$ for all $S_1, S_2 \subseteq \Gamma(v) \setminus \{w\}$.

With the notations of Proposition 1 denote by $\Psi^k_{(v,w)}$ the restriction of the map $\Psi^k_{(v,w)}$ to the $(k-1)$-element subsets of $\Gamma(v) \setminus \{w\}$. There are quite obvious algorithms that compute from the maps $\Psi^k_{(v,w)}$, $\{v, w\} \in E_P$, the k-faces of P (as subsets of V_P), and vice versa, in polynomial time in the number $f_k(P)$ of k-faces of the simple d-polytope P. Since both $f_2(P)$ as well as $f_{d-1}(P)$ are bounded polynomially in the size of G_P, the following result follows.

Corollary 1. *There are polynomial time algorithms that, given the graph G_P of a simple d-polytope P, compute the set of facets of P from the set of 2-faces of P (both viewed as sets of subsets of V_P), and vice versa.*

For the rest of this section let P_1 and P_2 be two simple polytopes, and let $g : V_{P_1} \longrightarrow V_{P_2}$ be an isomorphism of the graphs $G_{P_1} = (V_{P_1}, E_{P_1})$ and $G_{P_2} = (V_{P_2}, E_{P_2})$ of P_1 and P_2, respectively (i.e., g is an in both directions edge preserving bijection).

The core of Blind and Mani's paper [2] is the following result.

Proposition 2. *The graph isomorphism g maps every cycle in G_{P_1} that corresponds to a 2-face of P_1 to a cycle in G_{P_2} that corresponds to a 2-face of P_2.*

Blind and Mani's proof proceeds in the dual setting, i.e., in terms of the boundary complexes ∂P_1^* and ∂P_2^* of the simplicial dual polytopes P_1^* and P_2^* of P_1 and P_2, respectively. The strategy is to show that, if some cycle in G_{P_1} corresponding to a 2-face of P_1 was mapped to some cycle in G_{P_2} that does not correspond to any 2-face of P_2, then a certain sub-complex of ∂P_2^* would have a certain non-vanishing (reduced) homology group. They complete their proof of Proposition 2 by showing that the respective homology group, however, is zero. The key ingredient they use to prove this is the following. For each face F of P_1^* there is a *shelling order* of the facets of P_1^* (i.e., an ordering satisfying certain convenient topological properties, which, however, can be expressed completely combinatorially, see Sect. 3) in which the facets containing F come first.

From Proposition 2 one can deduce that the graph isomorphism g actually induces a bijection between the cycles in G_{P_1} that correspond to 2-faces of P_1 and the cycles in G_{P_2} that correspond to 2-faces of P_2. Once this is established, Proposition 1 yields the following result.

Theorem 1 (Blind & Mani [2]). *Every isomorphism between the graphs G_{P_1} and G_{P_2} of two simple polytopes P_1 and P_2, respectively, induces an isomorphism between the vertex-facet incidences of P_1 and P_2. In particular, the graph of a simple polytope determines its entire face lattice.*

3 Kalai's Constructive Proof

Kalai realized that the existence of shelling orders as exploited by Blind and Mani can be used directly in order to devise a simple proof which does not rely on any topological notions like homology [15]. He formulated his proof in the original setting, i.e., for simple polytopes, where the notion corresponding to "shelling" is called "abstract objective function."

From now on, let P be a simple d-polytope with n vertices. For simplicity of notation, we will identify each face of P not only with the corresponding subset of V_P, but also with the corresponding induced subgraph of G_P. Furthermore, by saying that $w \in W \subset V_P$ is a sink of W we mean that w is a sink of the orientation induced on the subgraph of G_P that is induced by W.

Definition 1. *Every bijection $\varphi : V_P \longrightarrow \{1, \ldots, n\}$ induces an acyclic orientation $\mathcal{O}_\varphi$ of the graph G_P of P, where an edge is directed from its larger end-node to its smaller end-node (with respect to φ). The map φ is called an* abstract objective function *(AOF) if $\mathcal{O}_\varphi$ has a unique sink in every non-empty face of P (including P itself). Such an orientation of G_P is called an AOF-orientation.*

The inverse orientation of an AOF-orientation is an AOF-orientation as well (this follows, e.g., from Theorem 3). Thus, every AOF-orientation also has a unique source in every non-empty face.

From the fact that the simplex algorithm works correctly (on every face) one easily derives that every linear function that assigns pairwise different values to the vertices of P induces an AOF-orientation (this is a consequence of the convexity of the faces). From this observation, the following fact follows (which is dual to the existence of the shelling orders required in Blind and Mani's proof).

Lemma 1. *Let $W \subset V_P$ be any face of P. There is an AOF-orientation of G_P for which W is* terminal, *i.e., no edge in the cut defined by W is directed from W to $V_P \setminus W$.*

In a sense, this statement can be reversed.

Lemma 2. *Let $W \subset V_P$ be a set of nodes inducing a k-regular connected subgraph of G_P, and let $\mathcal{O}$ be an AOF-orientation for which W is terminal. Then W is a k-face of P.*

Proof. Since $\mathcal{O}$ is acyclic, it has a source s in W. Let $w_1, \ldots, w_k \in W$ be the neighbors of s in W, and let $F := F(\{t, w_1, \ldots, w_k\}) \subset V_P$ be the k-face of P that is spanned by $t, w_1, \ldots, w_k$. Since $\mathcal{O}$ has unique sources on non-empty faces, $s \in W \cap F$ must be the unique source of F. By the acyclicity of $\mathcal{O}$ there hence is a monotone path from s to every node in F. Since W is terminal this implies $F \subseteq W$. Because both F and W induce k-regular connected subgraphs of G_P, $F = W$ follows.

Lemma 1 and Lemma 2 imply that one can compute the vertex-facet incidences of P, provided that one knows all AOF-orientations of G_P. Kalai's crucial discovery is that one can compute the AOF-orientations just from G_P (i.e., without explicitly knowing the faces of P).

Definition 2. *For an orientation $\mathcal{O}$ of G_P let $h_k(\mathcal{O})$ be the number of nodes with in-degree k. The number*

$$\mathcal{H}(\mathcal{O}) := \sum_{k=0}^{d} h_k(\mathcal{O}) \cdot 2^k$$

is called the $\mathcal{H}$-sum of $\mathcal{O}$.

Since every subset of neighbors of a vertex v of P together with v span a face of P containing no other neighbors of v, one finds (by double-counting) that

$$\mathcal{H}(\mathcal{O}) = \sum_{F \text{ face of } P} (\# \text{ sinks of } \mathcal{O} \text{ in } F) \tag{1}$$

is the total number of sinks induced by $\mathcal{O}$ on faces of P. Consequently, since every acyclic orientation has at least one sink in every non-empty face, we have the following characterization.

Lemma 3. *An orientation $\mathcal{O}$ of G_P is an AOF-orientation if and only if it is acyclic and has minimal $\mathcal{H}$-sum among all acyclic orientations of G_P (which then equals the number of non-empty faces of P).*

Thus, by enumerating all $2^{\frac{d \cdot n}{2}} = \sqrt{2}^{d \cdot n}$ orientations of G_P one can find all AOF-orientations of G_P.

Theorem 2 (Kalai [15]). *There is an algorithm that computes the vertex-facet incidences of a simple d-polytope with n vertices from its graph in $\mathcal{O}(\sqrt{2}^{d \cdot n})$ steps.*

4   Walks and Orientations

In this section, we refine the ideas of Kalai's proof and combine them with the observation (exploited by Blind and Mani) that it suffices to identify the 2-faces from the graph of a simple polytope, even with respect to the question for polynomial time reconstruction algorithms (see Corollary 1). Let us start with a result that emphasizes the importance of the 2-faces even more. The result was known for cubes [11]; for three-dimensional simple polytopes it was independently proved by Develin [4]). For general simple polytopes it seems that it was assumed to be false (see [19, Ex. 8.12 (iv)]).

Theorem 3. *An acyclic orientation $\mathcal{O}$ of the graph G_P of a simple polytope P is an AOF-orientation if and only if it has a unique sink on every 2-face of P.*

Proof. The "only if" part is clear by definition. For the "if" part, let $\varphi : V_P \longrightarrow \{1,\ldots,n\}$ be a bijection inducing an acyclic orientation $\mathcal{O} = \mathcal{O}_\varphi$ that has a unique sink in every 2-face of P. Suppose there is a face F of P in which $\mathcal{O}$ has two sinks $t_1, t_2 \in F$ ($t_1 \neq t_2$). We might assume $F = P$ (because F itself is a simple polytope with every 2-face of F being a 2-face of P).

Since G_P is connected, there is a path in G_P connecting t_1 and t_2. Let $\Pi \neq \varnothing$ be the set of all these paths. For every $\pi \in \Pi$ we denote by $\mu(\pi)$ the maximal φ-value of any node in π. Let $\pi_{\min} \in \Pi$ be a path with minimal μ-value among all paths in Π, where $v \in V_P$ is the node in $\pi_{\min}$ with $\varphi(v) = \mu(\pi_{\min})$ (see Fig. 2).

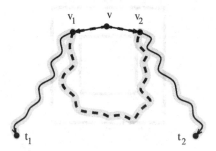

Fig. 2. Illustration of the proof of Theorem 3. The fat grey path is the one yielding the contradiction.

Obviously, v is a source in the path $\pi_{\min}$ (in particular, $v \notin \{t_1, t_2\}$). Let C be the 2-face spanned by v and its two neighbors v_1 and v_2 in $\pi_{\min}$. Since v is the unique source of $\mathcal{O}$ in C, v has the largest φ-value among all nodes in the union U of C and $\pi_{\min}$. But $U \setminus \{v\}$ induces a connected subgraph of G_P containing both t_1 and t_2, which contradicts the minimality of $\mu(\pi_{min})$.

From now on let, again, P be a simple polytope. The ultimate goal is to find the system of cycles in the graph G_P that corresponds to the set of 2-faces of P. However, we even do not know how to *prove* or *disprove* efficiently that a given system of cycles actually is the one we are searching for. We now define more general systems having the property that one can at least generate one of them in polynomial time (which in general, of course, will not be the desired one), and among which the one corresponding to the set of 2-faces of P can be characterized using AOF-orientations.

Definition 3. (i)   A sequence $W = (w_0, \ldots, w_{l-1})$ (with $l \geq 3$) of nodes in G_P is called a closed smooth walk in G_P, if $\{w_i, w_{i+1}\}$ is an edge of G_P and $w_{i-1} \neq w_{i+1}$ for all i (where, as in the following, all indices are taken modulo l). Note that the w_i need not be pairwise disjoint. We will identify two closed smooth walks if they differ only by a cyclic shift and/or a "reflection" of their node sequences.

(ii) A set $\mathcal{W}$ of closed smooth walks in G_P is a facoidal system of walks if for every triple $v, v_1, v_2 \in V_P$ ($v_1 \neq v_2$) such that both v_1 and v_2 are neighbors of v there is a unique closed smooth walk $(w_0, \ldots, w_{l-1}) \in \mathcal{W}$ with $(w_{i-1}, w_i, w_{i+1}) \in \{(v_1, v, v_2), (v_2, v, v_1)\}$ for some i, which is also required to be unique.

The system of 2-faces of P yields a uniquely determined (recall the identifications mentioned in part 1 of the definition) facoidal system of walks in G_P, which is denoted by $\mathcal{C}_P$. In general, there are many other facoidal systems of walks (see Fig. 3).

Fig. 3. A facoidal system of four walks in the graph of the three-dimensional cube.

For each path λ in G_P of length two denote by $v(\lambda)$ the inner node of λ. Let G_P^* be the graph defined on the paths of length two in G_P, where two paths λ_1 and λ_2 are adjacent if and only if they share a common edge and $v(\lambda_1) \neq v(\lambda_2)$ holds (see Fig. 4).

A *2-factor* in a graph G is a set of (not self-intersecting) cycles in G such that every node is contained in a unique cycle. Checking whether a graph has a 2-factor and finding one (if it exists) can be reduced (by a procedure due to Tutte [17]) to searching for a perfect matching in a related graph (which can be performed in polynomial time by Edmonds' algorithm [6]).

Fig. 4. The left constellation gives rise to an edge of $G_P^\star$, while the right one does not.

Proposition 3. *For simple polytopes P,*

(i) *there is a (polynomial time computable) bijection between the facoidal systems of walks in G_P and the 2-factors of $G_P^\star$,*

(ii) *checking whether a given set of node-sequences in G_P is a facoidal system of walks can be done in polynomial time, and*

(iii) *one can find a facoidal system of walks in G_P in polynomial time.*

Proof. Part 2 is obvious, part 3 follows from part 1 by Tutte's reduction [17] and Edmonds algorithm [6], and part 1 is readily obtained from the definitions.

Proposition 3 shows that facoidal systems of walks have quite convenient algorithmic properties. However, they become useful only due to the fact that the system $\mathcal{C}_P$ corresponding to the 2-faces of P can be well-characterized among them, as we will demonstrate next.

Definition 4. *Let $\mathcal{O}$ be any orientation of G_P.*

(i) *The $\mathcal{H}_2$-sum of $\mathcal{O}$ is defined as*

$$\mathcal{H}_2(\mathcal{O}) := \sum_{k=0}^{d} h_k(\mathcal{O}) \cdot \binom{k}{2} .$$

(ii) *A closed smooth walk $(w_0, \ldots, w_{l-1})$ in G_P has a* sink *(* source, respectively*) at position i (with respect to the orientation $\mathcal{O}$), if the edges $\{w_i, w_{i-1}\}$ and $\{w_i, w_{i+1}\}$ both are directed towards (away from, respectively) w_i.*

The following follows immediately from the definitions.

Lemma 4. *For every orientation $\mathcal{O}$ of G_P the sum $\mathcal{H}_2(\mathcal{O})$ equals the total number of sinks (with respect to $\mathcal{O}$) in every facoidal system of walks in G_P.*

Now we can formulate and prove the main result of this section (where $f_2(P)$ denotes the number of 2-faces of P).

Theorem 4. *Let P be a simple polytope, $\mathcal{W}$ a facoidal system of walks in G_P, and $\mathcal{O}$ an acyclic orientation of G_P. Then*

$$\#\mathcal{W} \leq f_2(P) \leq \mathcal{H}_2(\mathcal{O}) \tag{2}$$

holds.

(i) *The first inequality holds with equality if and only if $\mathcal{W} = \mathcal{C}_P$ (i.e., $\mathcal{W}$ "is" the set of 2-faces of P).*

(ii) *The second inequality holds with equality if and only if $\mathcal{O}$ is an AOF-orientation of G_P.*

Proof. Since $\mathcal{O}$ is acyclic, every closed smooth walk in G_P must have at least one sink with respect to $\mathcal{O}$. Thus, Lemma 4 implies

$$\#\mathcal{W} \leq \mathcal{H}_2(\mathcal{O}) \ , \tag{3}$$

yielding

$$f_2(P) = \#\mathcal{C}_P \leq \mathcal{H}_2(\mathcal{O}) \ , \tag{4}$$

where by Theorem 3 equality holds in (4) if and only if $\mathcal{O}$ is an AOF-orientation. Because G_P has an AOF-orientation $\mathcal{O}_0$ (see Lemma 1), inequality (3) gives $\#\mathcal{W} \leq \mathcal{H}_2(\mathcal{O}_0) = f_2(P)$.

Hence, it remains to prove that $\#\mathcal{W} = \#\mathcal{C}_P$ implies $\mathcal{W} = \mathcal{C}_P$. Suppose, that $\#\mathcal{W} = \#\mathcal{C}_P$ holds. It thus suffices to show $\mathcal{C}_P \subseteq \mathcal{W}$ (since we know already $\#\mathcal{C}_P \geq \#\mathcal{W}$). Let $C \in \mathcal{C}_P$ be any closed smooth walk corresponding to a 2-face of P. By Lemma 2 there is an AOF-orientation $\mathcal{O}_C$ of G_P such that C is terminal with respect to $\mathcal{O}_C$. Let $w_1 \in V_P$ be the unique source in C (with respect to $\mathcal{O}_C$), and let w_0 and w_2 be the two neighbors of w_1 in C. By definition, there is a (unique) $W = (w_0, w_1, w_2, \ldots, w_{l-1}) \in \mathcal{W}$. Because of $\#\mathcal{W} = \#\mathcal{C}_P = \mathcal{H}_2(\mathcal{O}_C)$ the closed smooth walk W has a unique sink at some position j and its unique source at position 1. Thus, the two paths $(w_1, w_2, \ldots, w_j)$ and $(w_1, w_0, \ldots, w_j)$ both are monotone. Since C is terminal, this implies that these two paths are contained in C. Therefore we have $C = W \in \mathcal{W}$.

5   Good Characterizations

Theorem 4 immediately yields characterizations of sets of 2-faces and of AOF-orientations that are similar to Kalai's characterization of AOF-orientations (see Lemma 3).

Corollary 2. *Let P be a simple polytope.*

(i) *A facoidal system of walks in G_P is the system $\mathcal{C}_P$ of 2-faces of P if and only if it has maximal cardinality among all facoidal systems of walks in G_P.*

(ii) *An acyclic orientation of G_P is an AOF-orientation if and only if it has minimal $\mathcal{H}_2$-sum among all acyclic orientations of G_P.*

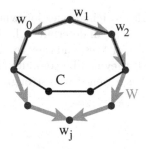

Fig. 5. Illustration of the proof of Theorem 4. If $W \neq C$ then C cannot be terminal.

Unfortunately, for arbitrary graphs the problem of finding a 2-factor with as many cycles as possible is $\mathcal{NP}$-hard. This follows from the fact that the question whether a graph can be partitioned into triangles is $\mathcal{NP}$-complete [7, Prob. GT11].

With respect to algorithmic questions, the following *good characterizations* (in the sense of Edmonds [5,6]) of the set of 2-faces (and thus, by Proposition 1 of the vertex-facet incidences) as well as of AOF-orientations may be more valuable than those in Corollary 2.

Corollary 3. *Let P be a simple polytope.*

(i) *Let $\mathcal{W}$ be a facoidal system of walks in G_P. Either there is an acyclic orientation of G_P having a unique sink in every walk of $\mathcal{W}$, or there is a facoidal system of walks in G_P of larger cardinality than $\#\mathcal{W}$. In the first case, $\mathcal{W} = \mathcal{C}_P$ "is" the set of 2-faces of P, in the second, it is not.*

(ii) *Let $\mathcal{O}$ be an acyclic orientation of G_P. Either there is a facoidal system $\mathcal{W}$ of walks in G_P such that $\mathcal{O}$ has a unique sink in every walk in $\mathcal{W}$, or there is an acyclic orientation of G_P with smaller $\mathcal{H}_2$-sum than $\mathcal{H}_2(\mathcal{O})$. In the first case, $\mathcal{O}$ is an AOF-orientation, in the second, it is not.*

For graphs G of simple polytopes let us define *Problem (A)* as

$$\max \#\mathcal{W} \quad \text{subject to} \quad \mathcal{W} \text{ facoidal system of walks in } G$$

and *Problem (B)* as

$$\min \mathcal{H}_2(\mathcal{O}) \quad \text{subject to} \quad \mathcal{O} \text{ acyclic orientation of } G \ .$$

A third consequence of Theorem 4 is the following result.

Corollary 4. *The Problems (A) and (B) form a pair of strongly dual combinatorial optimization problems. The optimal solution of Problem (A) yields the 2-faces of the respective polytope (and thus its vertex-facet incidences, see Proposition 1). Every optimal solution to Problem (B) is an AOF-orientation of the graph.*

Thus, the answer to Perles' original question whether the vertex-facet incidences of a simple polytope are at all determined by its graph, is not only "yes" (as proved by Blind and Mani), or "yes, and they can be computed" (as shown by Kalai), but at least "yes, and they can be computed by solving a combinatorial optimization problem that has a well-stated strongly dual problem."

6   Remarks

Corollary 4 suggests to design a primal-dual algorithm for the problem of reconstructing (the vertex-facet incidences of) a simple polytope from its graph. Such an algorithm would start by computing an arbitrary facoidal system $\mathcal{W}$ of walks in the given graph (see Proposition 3) and any acyclic orientation $\mathcal{O}$. Then it would check for $\#\mathcal{W} = \mathcal{H}_2(\mathcal{O})$. If equality holds then by Theorem 4 one is done. Otherwise, the algorithm would try to improve either $\mathcal{W}$ or $\mathcal{O}$ by exploiting the reasons for $\#\mathcal{W} \neq \mathcal{H}_2(\mathcal{O})$. For a concise treatment of different classical and recent applications of the primal-dual method in Combinatorial Optimization see [8].

Such a (polynomial time) primal-dual algorithm would in particular yield polynomial time algorithms for the problem to determine an (arbitrary) AOF-orientation from the graph G_P of a simple polytope P and the set of 2-faces of P, as well as for the problem to determine the 2-faces of P from G_P and an AOF-orientation. As for the first of these two problems it is worth to mention that no polynomial time method is known that would find any AOF-orientation even if the input is the entire face lattice of P. For the second problem no polynomial time algorithm is known as well.

Let (C) be the problem to decide for the graph G_P of a simple polytope P and a set $\mathcal{C}$ of subsets of nodes of G_P if $\mathcal{C}$ is the set of the subsets of nodes of G_P that correspond to the 2-faces of P. Let (D) be the problem to decide for the graph G_P of a simple polytope P and an orientation $\mathcal{O}$ of G_P if $\mathcal{O}$ is an AOF-orientation. The good characterizations in Corollary 3 may tempt one to conjecture that these two problems can be solved in polynomial time.

Unfortunately, from the complexity theoretic point of view, Corollary 3 does not provide us with any evidence for that. In particular, it does *not* imply that problems (C) and (D) are contained in $\mathcal{NP} \cap \text{co}\mathcal{NP}$. The reason is that the problem (G) to decide for a given graph if there is any simple polytope P such that G is isomorphic to the graph of P is neither known to be in $\mathcal{NP}$ nor in $\text{co}\mathcal{NP}$. Corollary 3 only shows that both problems (C) and (D) are in $\mathcal{NP} \cap \text{co}\mathcal{NP}$, if one restricts them to any class of graphs for which problem (G) is in $\mathcal{NP} \cap \text{co}\mathcal{NP}$.

The problem (S) to decide for a given lattice L if there is a simple polytope P such that the face lattice L_P of P is isomorphic to L is known as the *Steinitz problem for simple polytopes*.

Corollary 5. *If problem (G) is contained in $\mathcal{NP}$ or $\text{co}\mathcal{NP}$, then problem (S) is contained in $\mathcal{NP}$ or $\text{co}\mathcal{NP}$, respectively.*

Proof. The face lattice L_P of a polytope P is ranked. The graph G having the rank one elements of L_P as its nodes, where two rank one elements are adjacent if and only if they are below a common rank two element, is isomorphic to the graph of P. It can be computed from L_P in polynomial time. Corollary 1 shows that one can compute the vertex-facet incidences (and thus the entire face lattice L_P, see the first paragraph of the introduction) of a simple polytope P in polynomial time (in the size of L_P) from the poset that is induced by the elements of rank one (corresponding to the vertices), rank two (corresponding to the 1-faces), and rank three (corresponding to the 2-faces). Together with the first part of Corollary 3 this proves the claim.

Extending results of Mnëv [16] by using techniques described in [3] one finds (see [1, Cor. 9.5.11]) that there is a polynomial (Karp-)reduction of the problem to decide whether a system of linear inequalities has an integral solution to problem (S). Thus, problem (S) (and therefore, by Corollary 5, problem (G)) is not contained in $co\mathcal{NP}$, unless $\mathcal{NP} = co\mathcal{NP}$. Furthermore, there are rational simple polytopes P with the property that every rational simple polytope Q whose graph is isomorphic to the graph of P has vertices with super-polynomial coding lengths in the size of the graphs (this follows from Theorem B in [9]). Thus, it seems also unlikely that problem (G) is contained in $\mathcal{NP}$.

The results presented in Sect. 4 hence do neither lead to efficient algorithms nor to new examples of problems in $\mathcal{NP} \cap co\mathcal{NP}$ not (yet) known to be solvable in polynomial time. Nevertheless, they show that the problem to reconstruct a simple polytope from its graph can be modeled as a combinatorial optimization problem with a strongly dual problem. We hope that this is an appearance of Combinatorial Optimization Jack Edmonds is pleased to see in this volume dedicated to him.

Acknowledgements. I thank Günter M. Ziegler for valuable comments on an earlier version of the manuscript.

References

1. A. Björner, M. Las Vergnas, B. Sturmfels, N. White, and G. M. Ziegler. *Oriented Matroids (2nd ed.)*, volume 46 of *Encyclopedia of Mathematics and Its Applications*. Cambridge University Press, 1999.
2. R. Blind and P. Mani-Levitska. Puzzles and polytope isomorphisms. *Aequationes Math.*, 34:287–297, 1987.
3. J. Bokowski and B. Sturmfels. *Computational Synthetic Geometry*, volume 1355 of *Lecture Notes in Mathematics*. Springer, Heidelberg, 1989.
4. M. Develin. E-mail conversation, Nov 2000. develin@bantha.org.
5. J. Edmonds. Maximum matching and a polyhedron with 0,1-vertices. *J. Res. Natl. Bur. Stand. – B (Math. and Math. Phys.)*, 69B:125–130, 1965.
6. J. Edmonds. Paths, trees, and flowers. *Can. J. Math.*, 17:449–467, 1965.
7. M. R. Garey and D. S. Johnson. *Computers and Intractability. A Guide to the Theory of NP-Completeness*. W. H. Freeman and Company, New York, 1979.

8. M. X. Goemans and D. P. Williamson. The primal-dual method for approximation algorithms and its application to network design problems. In D. Hochbaum, editor, *Approximation Algorithms*, chapter 4. PWS Publishing Company, 1997.

9. J. E. Goodman, R. Pollack, and B. Sturmfels. The intrinsic spread of a configuration in $\mathbb{R}^d$. *J. Am. Math. Soc.*, 3(3):639–651, 1990.

10. C. Haase and G. M. Ziegler. Examples and counterexamples for Perles' conjecture. Technical report, TU Berlin, 2001. To appear in: Discrete Comput. Geometry.

11. P. L. Hammer, B. Simeone, T. M. Liebling, and D. de Werra. From linear separability to unimodality: A hierarchy of pseudo-boolean functions. *SIAM J. Discrete Math.*, 1:174–184, 1988.

12. M. Joswig, V. Kaibel, and F. Körner. On the k-systems of a simple polytope. Technical report, TU Berlin, 2001. arXiv: math.CO/0012204, to appear in: Israel J. Math.

13. M. Joswig and G.M. Ziegler. Neighborly cubical polytopes. *Discrete Comput. Geometry*, 24:325–344, 2000.

14. V. Kaibel and M. Pfetsch. Computing the face lattice of a polytope from its vertex-facet incidences. Technical report, TU Berlin, 2001. arXiv:math.MG/01060043, submitted.

15. G. Kalai. A simple way to tell a simple polytope from its graph. *J. Comb. Theory, Ser. A*, 49(2):381–383, 1988.

16. N. E. Mnëv. The universality theorems on the classification problem of configuration varieties and convex polytopes varieties. In O.Ya. Viro, editor, *Topology and Geometry – Rohlin Seminar*, volume 1346 of *Lecture Notes in Mathematics*, pages 527–543. Springer, Heidelberg, 1988.

17. W. T. Tutte. A short proof of the factor theorem for finite graphs. *Can. J. Math.*, 6:347–352, 1954.

18. H. Whitney. Congruent graphs and the connectivity of graphs. *Am. J. Math.*, 54:150–168, 1932.

19. G. M. Ziegler. *Lectures on Polytopes*. Springer-Verlag, New York, 1995. Revised edition 1998.

An Augment-and-Branch-and-Cut Framework for Mixed 0-1 Programming

Adam N. Letchford[1] and Andrea Lodi[2]

[1] Department of Management Science, Lancaster University,
Lancaster LA1 4YW, England
A.N.Letchford@lancaster.ac.uk
[2] DEIS, University of Bologna, Viale Risorgimento 2, 40136 Bologna, Italy
alodi@deis.unibo.it

Abstract. In recent years the *branch-and-cut* method, a synthesis of the classical *branch-and-bound* and *cutting plane* methods, has proven to be a highly successful approach to solving large-scale integer programs to optimality. This is especially true for *mixed 0-1* and *pure 0-1* problems. However, other approaches to integer programming are possible. One alternative is provided by so-called *augmentation* algorithms, in which a feasible integer solution is iteratively improved (augmented) until no further improvement is possible.

Recently, Weismantel suggested that these two approaches could be combined in some way, to yield an *augment-and-branch-and-cut* (ABC) algorithm for integer programming. In this paper we describe a possible implementation of such a finite ABC algorithm for mixed 0-1 and pure 0-1 programs. The algorithm differs from standard branch-and-cut in several important ways. In particular, the terms *separation*, *branching*, and *fathoming* take on new meanings in the primal context.

1 Introduction

One of the most successful methods for solving Integer and Mixed-Integer Linear Programs (ILPs and MILPs, respectively) is the *branch-and-cut* approach, in which strong cutting planes are used to strengthen the linear programming relaxations at each node of a branch-and-bound tree (see Padberg & Rinaldi [22]; Caprara & Fischetti [5]). Branch-and-cut is currently very popular because it appears to be much more robust than the use of either cutting planes or branching in isolation.

However, although branch-and-cut is popular, research has been and still is being conducted into other approaches to integer programming — based on Lagrangian, surrogate or group relaxation, lattice basis reduction, test sets, and so on (see, e.g., Nemhauser & Wolsey [19]). Of particular interest in the present paper are so-called *augmentation* algorithms, in which a feasible solution is iteratively improved (augmented) until no further improvement is possible (and it can be proved that this is the case). Some recent results on augmentation algorithms can be found in Firla et al. [8], Haus et al. [14], Schulz et al. [23], Thomas [24] and Urbaniak et al. [25].

M. Jünger et al. (Eds.): Combinatorial Optimization (Edmonds Festschrift), LNCS 2570, pp. 119–133, 2003.
© Springer-Verlag Berlin Heidelberg 2003

Recently, Weismantel [26] suggested the possibility of somehow combining elements of augmentation and branch-and-cut algorithms, to yield an *augment-and-branch-and-cut* algorithm for integer programming — with the convenient acronym ABC. In this paper we describe such an ABC algorithm for *mixed 0-1* programs. It is based on a new primal cutting plane algorithm which we presented in [17], combined with some *branching rules* which are specifically tailored to the primal context.

The remainder of the paper is structured as follows. In Section 2 we briefly review the literature on branch-and-cut methods. In Section 3 we review the literature on primal cutting plane methods, including our new algorithm. In Section 4, we examine the issue of *branching* in the primal context. In Section 5 we show how the various components of the algorithm are integrated to give the general-purpose ABC method for mixed 0-1 programs. Preliminary computational results are reported in Section 6, while conclusions are given in Section 7.

2   Basic Concepts of Branch-and-Cut

In this section we briefly review the literature on standard branch-and-cut algorithms. Although much of this material is widely known, it is necessary to include it here so that in subsequent sections we can show how our primal ABC approach departs from the standard approach.

Suppose that a MILP has n integer-constrained variables, p continuous variables and m linear constraints. The vector of integer-constrained variables will be denoted by x and the vector of continuous variables by y. Let us suppose that the MILP takes the form:

$$\max\{c^T x + d^T y : Ax + Gy \leq b, x \in \mathbb{Z}_+^n, y \in \mathbb{R}_+^p\},$$

where c and d are objective coefficient vectors, A and G are matrices of appropriate dimension ($m \times n$ and $m \times p$, respectively) and b is an m-vector of right hand sides.

The feasible region of the *linear programming* (LP) *relaxation* of the MILP is the polyhedron

$$P := \{(x, y) \in \mathbb{R}_+^{n+p} : Ax + Gy \leq b\},$$

and the convex hull of feasible MILP solutions is the polyhedron

$$P_I := \text{conv}\{x \in \mathbb{Z}_+^n, y \in \mathbb{R}_+^p : Ax + Gy \leq b\}.$$

We have $P_I \subseteq P$ and in this paper we assume that containment is strict. For simplicity we also assume throughout the paper that the MILP is a *maximization* problem.

Suppose that the vector (x^*, y^*) is an optimal solution to the LP relaxation. If x^* is integral, then (x^*, y^*) is an optimal solution to the MILP and we are done. If not, then the value of the objective function provides an upper bound on the value of the optimum, but further work will be needed to solve the MILP to optimality. We can either *cut* (add extra inequalities) or *branch* (divide the problem into subproblems).

2.1   Cutting

A *cutting plane* (or *cut*) is an extra linear inequality which (x^*, y^*) does not satisfy, but which is satisfied by all solutions to the MILP. That is, the inequality must be valid for P_I but not for P. If we can find a cut, then we can add it to the LP relaxation and re-solve (typically via the dual simplex method), to obtain a new (x^*, y^*). If the new x^* is integral, we are done, otherwise the procedure of adding cuts continues until x^* becomes integral. As cuts are added, we obtain a non-increasing sequence of upper bounds.

In order to generate a cutting plane, one faces the following problem:

The Separation Problem: *Given some $(x^*, y^*) \in P$, find an inequality which is valid for P_I and violated by (x^*, y^*), or prove that none exists.*

A famous theorem of Grötschel, Lovász & Schrijver [13] states that, under certain technical assumptions, the separation problem is $\mathcal{NP}$-hard if and only if the original MILP is. However, if (x^*, y^*) is an extreme point of the current LP relaxation, which will be the case if the simplex method is being used, then cuts can be generated fairly easily. (For example, one can use the cuts of Gomory [11], [12]; or the disjunctive cuts of Balas, Ceria & Cornuéjols [1].)

In general, cutting plane algorithms based on 'general-purpose' cuts such as Gomory or disjunctive cuts exhibit slow convergence. This can be alleviated somewhat by adding several cuts in one go before reoptimizing by dual simplex, see for example Balas, Ceria & Cornuéjols [1] and Balas, Ceria, Cornuéjols & Natraj [3]. In general, however, it is preferable to use inequalities which take problem structure into account, especially inequalities which are 'deep' in the sense of inducing facets (or faces of high dimension) of the polyhedron P_I (see Padberg & Grötschel [20]; Nemhauser & Wolsey [19]).

It is often the case that several classes of deep inequalities are known for a given problem, and frequently each class contains an exponential number of members. To use a particular class of inequalities in practice, one needs to solve the following modified separation problem:

The Separation Problem for a Class of Inequalities: *Given some class $\mathcal{F}$ of inequalities which are valid for P_I and some $(x^*, y^*) \in P$, find a member of $\mathcal{F}$ which is violated by (x^*, y^*), or prove that none exists.*

It frequently happens that this modified separation problem is polynomially solvable for some classes of inequalities, and $\mathcal{NP}$-hard for others. Yet, even in the latter case it is frequently possible to devise heuristics for separation which perform reasonably well in practice.

Cutting plane algorithms based on deep inequalities typically converge much more quickly than algorithms based on general-purpose cuts. However, there is one disadvantage: due to the heuristic nature of the separation algorithms, or due to the lack of a complete description of P_I, it may happen that no deep cuts can be found even though x^* is still fractional. (This never happens with

Gomory or disjunctive cuts.) If we want to solve the MILP to optimality, and if we want to avoid using general-purpose cuts, then we will need to branch as described in the next subsection (or use some other solution technique).

Note that the cutting plane procedure yields a (typically good) upper bound on the optimum.

2.2   Branching

Instead of adding cuts to remove an invalid vector (x^*, y^*), one can *branch* instead, i.e., divide the original problem into subproblems. The most common method of branching is to choose an index i such that x_i^* is fractional and to create two subproblems. In one, the constraint $x_i \le \lfloor x_i^* \rfloor$ is added; in the other, the constraint $x_i \ge \lceil x_i^* \rceil$ is added. (Here, $\lfloor \cdot \rfloor$ and $\lceil \cdot \rceil$ denote rounding down and rounding up to the nearest integer, respectively.) In this way, x^* is excluded from each of the two branches created. Each of these subproblems can be quickly solved using the dual simplex method.

In the standard *branch-and-bound* method, often attributed to Land & Doig [15], recursive branching leads to a tree-like structure of subproblems. (In the case of mixed 0-1 programs, for example, a given node of the branch-and-bound tree will correspond to fixing a specified subset of the variables to zero and another specified subset of the variables to one.) The tree is then explored, e.g., by breadth-first or depth-first search. Along the way, branches are 'pruned' (removed from consideration) when their associated LP upper bound is lower than the current lower bound (corresponding to the best integer solution found so far). The algorithm terminates when the only remaining node is the root node of the tree.

Note that in general the set of variables for which x_i^* is fractional may be large, and therefore some kind of heuristic rule is needed for choosing the branching variable. A common method is to choose the variable whose fractional part is closest to one-half, but other rules can perform better, particularly if they take specific problem structure into account.

2.3   The Overall Scheme

The two methods of cutting and branching are in a sense complementary. Cutting planes, especially deep ones, can lead to very tight upper bounds, but may not yield a feasible solution for a long time. Branching, on the other hand, allows one to find feasible solutions relatively quickly, but the upper bounds may be too weak to limit the size of the search tree.

The natural solution is to integrate cutting and branching within a single algorithm. This yields the *branch-and-cut* technique, in which cutting planes are used at *each node* of the branch-and-bound tree to strengthen the LP relaxations (see Padberg & Rinaldi [22]; Balas, Ceria & Cornuéjols [2]; Balas, Ceria, Cornuéjols & Natraj [3]; Caprara & Fischetti [5]).

In branch-and-cut it is normal to use inequalities which are valid for P_I as cutting planes. Inequalities of this kind are valid *globally*, i.e., at every node of the branch-and-cut tree. Thus, any violated inequality generated at any node may be used to strengthen the upper bound at any other node of the tree (assuming of course that it is violated). This would not be possible with cuts which are only valid at a particular node of the tree.

There are a few more ingredients which are needed to form a sophisticated branch-and-cut algorithm. We have already mentioned rules for selecting the branching variable, but there are other key components such as pricing, variable fixing, cut pools, and so on (Padberg & Rinaldi [22]). In Section 5 we briefly review these and consider how to adapt them to the primal context.

3   Primal Cutting Plane Algorithms

3.1   The Basic Concept

The goal of a primal cutting plane algorithm is to iteratively move from one corner of P_I to a better one until no further improvement is possible. To begin the method, one must know of a feasible solution to the ILP, preferably a good one, which we will denote by $(\bar{x}, \bar{y})$. Moreover, it must be a solution with a very specific property: it must be an extreme point of both P and P_I. Equivalently, it must be a basic feasible solution to the LP relaxation of the problem.

If no such $(\bar{x}, \bar{y})$ is known, then artificial variables must be used to find one, much as in the ordinary simplex method. However, in practice it is often easy to find one. Indeed, for mixed 0-1 problems, all we need to do is find *any* feasible solution to the MILP: it can be converted into a basic feasible solution by fixing the x variables at their given values and solving an LP to determine the y variables.

Given a suitable $(\bar{x}, \bar{y})$, one then constructs the associated simplex tableau. If $(\bar{x}, \bar{y})$ is dual feasible, it is optimal and the method stops. If not, a primal simplex pivot is made, leading to a new vector (or the same one in the case of degeneracy), which we will denote by (x^*, y^*). If x^* is integral, then we have found an improved MILP solution, and (x^*, y^*) becomes the new $(\bar{x}, \bar{y})$. If on the other hand it is fractional, then a cutting plane is generated which cuts off (x^*, y^*). Then, another attempt is made to pivot from $(\bar{x}, \bar{y})$, leading to a different (x^*, y^*), and so on. The method terminates when $(\bar{x}, \bar{y})$ has been proved to be dual feasible.

In fact, it is possible to compute (x^*, y^*) from the information in the $(\bar{x}, \bar{y})$ tableau without actually performing a pivot. Therefore it is not actually necessary to pivot to (x^*, y^*) in order to determine if it is a feasible MILP solution or not. For this reason, some primal cutting plane algorithms only perform the pivot explicitly when it leads to an augmentation.

3.2   Algorithms from the 1960s

Several primal cutting plane algorithms, for pure ILPs only, appeared in the 1960s (Ben-Israel and Charnes [4], Young [27], [28], Glover [10]). The algorithms in [4] and [27] are extremely complicated and, as shown in [28] and [10], they can be simplified considerably without affecting convergence. The simplest of these algorithms, and in our view the best one, is that of Young [28].

Detailed descriptions of the Young algorithm can be found in Garfinkel & Nemhauser [9], in our paper [17], and also in Firla et al. [8]. The basic idea behind Young's algorithm is to begin with an all-integer tableau and to perform primal simplex pivots whenever the pivot element is equal to 1. In this way one guarantees that the subsequent tableau (and the associated x^*) is also all-integer. If the pivot element is not equal to 1, then a cutting plane is added to the tableau so that the pivot element in the enlarged tableau *is* 1.

Young proved that, with an appropriate lexicographic variable selection rule, his algorithm is finitely convergent. However, the performance of the algorithm in practical computation has been disappointing. The main problem is that one often encounters extremely long sequences of degenerate pivots, without either augmenting or proving dual feasibility.

3.3   Padberg and Hong's Algorithm

The key to obtaining a viable primal cutting plane algorithm is the use of strong (preferably facet-inducing) cutting planes. To our knowledge the first authors to use strong cutting planes in a primal context were Padberg & Hong [21]. (A detailed description of this paper appears in Padberg & Grötschel [20] and in our paper [17].) Padberg and Hong implemented a primal cutting plane algorithm for the *Travelling Salesman Problem* (TSP), based on facet-defining cuts such as *subtour elimination constraints* (SECs) and *2-matching, comb* and *chain* inequalities.

The algorithms used by Padberg and Hong to generate violated inequalities — which they called *constraint identification algorithms* — are very similar to what are now known as separation algorithms. However, there was a subtle difference: as well as requiring the inequality to be violated by x^*, they also required that it be *tight* (satisfied as an equation) at $\bar{x}$. The reason for this is that, in the case of 0-1 problems like the TSP, only inequalities which are tight at $\bar{x}$ can help in either augmenting or proving dual feasibility of $\bar{x}$.

Therefore the problem solved by Padberg and Hong's algorithms is as follows (see also [17], [16], [7]):

The Primal Separation Problem for a Class of Inequalities:
Given some class $\mathcal{F}$ of inequalities which are valid for P_I, some $(x^*, y^*) \in P$ and some $(\bar{x}, \bar{y})$ which is an extreme point of both P and P_I, find a member of $\mathcal{F}$ which is violated by (x^*, y^*) and tight for $(\bar{x}, \bar{y})$, or prove that none exists.

It is easy to show (see Padberg & Grötschel [20]) that a given primal separation can be transformed to the equivalent standard (dual) version. However, the reverse does not hold in general and in the papers Letchford & Lodi [16] and Eisenbrand, Rinaldi & Ventura [7] it is shown that for many classes of inequalities, primal separation is substantially *easier* than standard separation. This gives some encouragement for pursuing the primal approach, especially for (mixed) 0-1 problems.

3.4 Our Algorithm

In our recent paper [17] we argued that primal cutting plane algorithms were worthy of more attention, and we proposed a generic algorithm for the case of mixed 0-1 problems based on primal separation algorithms. For the sake of brevity we do not describe this in detail here, but the basic structure of the algorithm is as follows:

- Step 1: Find a 'good' initial basic feasible solution $(\bar{x}, \bar{y})$ and construct an appropriate tableau.
- Step 2: If $(\bar{x}, \bar{y})$ is dual feasible, stop.
- Step 3: Perform a primal simplex pivot. If it is degenerate, return to step 2.
- Step 4: Let (x^*, y^*) be the new vector obtained. If (x^*, y^*) is a feasible solution to the MILP, set $(\bar{x}, \bar{y}) := (x^*, y^*)$ and return to step 2.
- Step 5: Call primal separation for known strong inequalities. If any are found, pivot back to $(\bar{x}, \bar{y})$, add one or more cuts to the LP, and return to step 3.
- Step 6: If x^* is integral but (x^*, y^*) is infeasible, call the 'special' cut generating procedure (described below) and return to step 3.
- Step 7: Generate one or more general-purpose cuts (such as Gomory fractional or mixed-integer cuts), add them to the LP, pivot back to $(\bar{x}, \bar{y})$ and return to step 3.

The reason that step 6 is necessary is that we do not require that the entire constraint system $Ax + Gy \leq b$ be present in the initial LP. (For some problems, the constraint system is too large to be handled all at once.) Therefore it is theoretically possible to obtain a vector (x^*, y^*) which has an integral x component, but which is not a feasible MILP solution because it violates a constraint in the system $Ax + Gy \leq b$ which is *not tight* at $(\bar{x}, \bar{y})$.

The 'special' cut generating procedure to handle this exceptional case is as follows:

- 6.1. Find an inequality in the system $Ax + Gy \leq b$ which is violated by (x^*, y^*) yet *not tight* at $(\bar{x}, \bar{y})$, and add it to the LP.
- 6.2. Perform a dual simplex pivot to arrive at a new point $(\hat{x}, \hat{y})$ which is a convex combination of $(\bar{x}, \bar{y})$ and (x^*, y^*). Set $(x^*, y^*) := (\hat{x}, \hat{y})$.
- 6.3. Generate a Gomory mixed-integer cut from a row of the tableau which corresponds to a fractional structural variable and add it to the LP. (We show in [17] that it is guaranteed to be tight at $(\bar{x}, \bar{y})$.)
- 6.4. Perform a dual simplex pivot to return to $(\bar{x}, \bar{y})$ and discard the inequality generated in step 6.2 (which is no longer tight).

In a sense, the mixed-integer cut generated in step 6.3 is a 'rotated' version of the inequality generated in step 6.1 — rotated in such a way as to provide a solution to the primal separation problem.

Computational results given in [17] clearly demonstrate that this algorithm is significantly better than that of Young. Nevertheless, it seems desirable to branch in step 7 instead of adding general-purpose cuts. That is our goal in the next section.

4   Branching in a Primal Context

Branching is desirable in ordinary branch-and-cut when (x^*, y^*) has a fractional x component, but the separation algorithms fail to find any violated cuts. In our primal approach, branching is desirable when $(\bar{x}, \bar{y})$ is not dual feasible, the adjacent point (x^*, y^*) has a fractional x component, and the *primal* separation algorithms fail.

However, branching is problematic in the primal context. Suppose one tried to branch in the standard way, by picking a variable index i such that x_i^* is fractional, and imposing either $x_i = 0$ or $x_i = 1$. Then the current best MILP solution $(\bar{x}, \bar{y})$ would be excluded from one of the two subproblems, and we would lose our starting basis on one of the two branches. Therefore, just as a non-standard separation problem appears in the primal context, a non-standard form of branching is also needed. One wants to branch in such a way that (x^*, y^*) is removed, but $(\bar{x}, \bar{y})$ remains intact on both branches.

We have developed suitable branching rules for 0-1 MILPs, which we now describe.

First, let us assume for simplicity of notation that the 0-1 variables have been complemented so that $\bar{x}$ is a vector of n zeroes, which we denote here by $\mathbf{0}$. Moreover, let us also assume that to save on memory our LP contains only constraints which are tight for $(\bar{x}, \bar{y})$, together with upper bounds of 1 on the x variables. (Our cutting plane algorithm described in Subsection 3.4 is designed to run in this way.) We begin with a fairly naive branching rule:

> **Naive Branching Rule:** *Suppose that there are a pair of variable indices i and j such that $0 < x_i^* < x_j^* \leq 1$. Create two branches, one with the constraint $x_i = 0$ added; the other with the constraint $x_i \geq x_j$ added.*

With this rule, no feasible solution is lost, $(\bar{x}, \bar{y})$ is feasible for both branches, and (x^*, y^*) is removed in both branches. One drawback however is that any feasible solution (x, y) which satisfies $x_i = x_j = 0$ will be valid on both branches. It would be more desirable to branch in such a way that no feasible solution, apart from $(\bar{x}, \bar{y})$ itself, appeared on both branches.

Let us consider the issue in more detail. The goal of branching is to force a variable x_i to be integral, even though x_i^* is currently fractional. An ideal branching rule would therefore be $(x_i = 0) \vee (x_i = 1)$, but, as we have seen, this would make $(\bar{x}, \bar{y})$ infeasible on the 'up'-branch. So, suppose that the current

LP feasible region (including any cutting planes added) is of the form: $\{x \in [0,1]^n, y \in \mathbb{R}_+^p : \bar{A}x + \bar{G}y \leq \bar{b}\}$, where all of the inequalities in the system $\bar{A}x + \bar{G}y \leq \bar{b}$ are currently tight for $(\bar{x}, \bar{y})$. If we *had* imposed $x_i = 1$, then the resulting feasible region would be

$$P^1 := \{x \in [0,1]^n, y \in \mathbb{R}_+^p : \bar{A}x + \bar{G}y \leq \bar{b}, x_i = 1\}.$$

Given that we do not want to remove $(\bar{x}, \bar{y})$, we consider instead the smallest polyhedron containing both $(\bar{x}, \bar{y})$ and P^1. It is not difficult to show that, when $\bar{x} = \mathbf{0}$, this polyhedron is:

$$P^2 := \text{conv}\{\mathbf{0} \cup P^1\} = \{x \in [0,1]^n, y \in \mathbb{R}_+^p : \bar{A}x + \bar{G}y \leq \bar{b}, x_j \leq x_i \ \forall j \neq i\}.$$

By definition, this new polyhedron has no extreme points in which x_i is fractional. Moreover, the next time we attempt to pivot from $(\bar{x}, \bar{y})$, we will arrive at a point (x^*, y^*) with $x_i^* = 1$.

Therefore our main branching rule for 0-1 MILPs is as follows:

Main Branching Rule: *Suppose that $0 < x_i^* < 1$. Create two branches, one with the constraint $x_i = 0$ added; the other with the constraints $x_j \leq x_i \ \forall j \neq i$ added.*

Note that $(\bar{x}, \bar{y})$ remains basic after the change.

With this branching rule, just as with the original one, no feasible solution is lost, $(\bar{x}, \bar{y})$ is feasible for both branches, and (x^*, y^*) is removed in both branches. However, in addition, the only vectors which can appear on both branches are those with x component equal to $\mathbf{0}$. This is therefore a more powerful partition, which we would expect to lead to quicker convergence of the algorithm.

Let us consider what happens when several branchings have occurred. Suppose that we wish to 'fix' variables x_i for $i \in N_0$ to zero, and variables x_i for $i \in N_1$ to one. The polyhedron of interest is now the convex hull of $(\bar{x}, \bar{y})$ and the set $\{x \in [0,1]^n, y \in \mathbb{R}_+^p : \bar{A}x + \bar{G}y \leq \bar{b}, x_i = 0 \ (i \in N_0), x_i = 1 \ (i \in N_1)\}$. By a similar argument to that given above it can be shown that, when $\bar{x} = \mathbf{0}$, the polyhedron in question is

$$\begin{aligned}
P^3 := \{x \in [0,1]^n, y \in \mathbb{R}_+^p : \ & \bar{A}x + \bar{G}y \leq \bar{b}, \\
& x_j = 0 \ \forall j \in N_0, \\
& x_j = x_i \ \forall j \in N_1 \setminus \{i\}, \\
& x_j \leq x_i \ \forall j \notin (N_0 \cup N_1)\},
\end{aligned} \tag{1}$$

where i is an arbitrary index in N_1.

That is, in order to perform further branching in an 'upward' direction it is merely necessary to add equations of the form $x_j = x_i$. Thus the system of inequalities can be easily modified as branching progresses.

Note that the very compact and clean form of polyhedron P^3 is possible because all the constraints in the LP before the branching operation are tight at

$(\bar{x}, \bar{y})$. If it is desired to include non-tight constraints in the LP, then it is still possible to perform the above branching but the non-tight inequalities require some additional 'work'.

Note also that the same kind of argument does not apply to general MILPs (i.e., MILPs in which the integer variables are not restricted to be binary). The branching rule implicitly relies on the fact that any 0-1 vector $\bar{x}$ is an extreme point of the unit hypercube. It is this property which enables us to complement variables in order to get $\bar{x} = \mathbf{0}$. There is no analogous complementation procedure in the case of general MILPs.

To close this section, we must mention one potential problem. If only constraints which are tight are included in the LP (along with the upper bounds on the x variables), then there is a (small) risk that the LP will be *unbounded*. If this happens, however, it will be because the profit can be increased without limit by changing $\bar{y}$ while leaving $\bar{x}$ unchanged. The solution to this is to solve a (typically small) LP to see if it is possible to augment by changing only the y component. If so, then one should augment and continue from there.

5   The Overall ABC Algorithm

At this point we have the main ingredients for the ABC algorithm: the primal separation component and the branching rules. However, some more details are necessary in order to specify how the overall algorithm works.

Fathoming of Nodes: Obviously we need some way of pruning the branching tree and, in particular, of fathoming a node. It is not difficult to see, from the properties of the primal simplex method, that in ABC a node can be fathomed when $(\bar{x}, \bar{y})$ is dual feasible.

Cut Pool: In order to keep the size of the basis small, it is normal in standard branch-and-cut to delete inequalities from the LP whenever their slack exceeds some small positive value. However, to avoid wasting time by separating the same inequality more than once, it is common to store these constraints in a so-called *cut pool* (e.g., Padberg & Rinaldi [22]). The natural primal analogue of this is as follows: tight cuts are kept in the LP and non-tight cuts are kept in the pool. Whenever $(\bar{x}, \bar{y})$ is augmented, constraints which are no longer tight can be put into the pool and any constraints in the pool which have become tight can be put into the LP.

Handling Augmentations: At first sight it would appear that every time an improved feasible solution $(\bar{x}, \bar{y})$ is found, it will be necessary to discard the branch-and-cut tree and begin branching and cutting from scratch. In fact, this is *not* necessary. It is possible to work with a single tree. When a node is fathomed, it means that no feasible solution exists with objective value greater than $(\bar{x}, \bar{y})$ when the associated variables are fixed. Given that the new $(\bar{x}, \bar{y})$ has a greater objective value than the old one, this remains true after the augmentation. Hence, it is necessary only to construct a new basis at the root node, which can be done using the cuts which are now tight at the new $(\bar{x}, \bar{y})$.

Pricing: When n is very large, it is normal practice in standard branch-and-cut to begin with only a subset of the variables and to include other variables only when needed. This is done by *pricing*, i.e., computing the LP reduced costs of the remaining variables and adding the variables whose reduced costs are positive (Padberg & Rinaldi [22]). This can be done in the primal approach as well.

Handling Problems with m Huge: As mentioned, for many important problems, the number of constraints m needed to define P is exponential in the problem input, but optimization over P is still possible because the (standard) separation problem for these constraints is solvable in polynomial time. These problems can be dealt with in the primal context also, because (as explained in Section 4) we only keep tight constraints in the LP.

Finally, the reader will have noticed that up to now there has been no mention of upper bounds in the ABC context. This is because, strictly speaking, they are not needed: to fathom a node of the tree, it is only necessary to prove dual feasibility. Nevertheless, there are reasons for thinking that some kind of upper bounding mechanism might be desirable. The main one is this: if for some reason the ABC algorithm has to be interrupted before optimality has been achieved, then an upper bound can be used to assess the quality of the final $(\bar{x}, \bar{y})$.

The simplest way to produce an upper bound, based on the idea of Padberg and Hong, is to solve the final LP at the root node to optimality. This LP can be solved in a relatively small number of primal simplex pivots, because $(\bar{x}, \bar{y})$ can be used as a starting basis. However, note that the resulting upper bound is unlikely to be better than the upper bound which would be obtained from a dual approach (assuming that similar inequalities and separation algorithms are used).

Another reason for wanting an upper bounding mechanism is to somehow eliminate variables from the problem entirely. In standard branch-and-cut, this is done as follows. Any variable with a reduced cost greater than the difference between the current upper and lower bounds may be eliminated from the problem (i.e., fixed at zero). This is called *reduced cost fixing*. In the ABC context we can do something similar, at least at the root node, by using the reduced costs from the optimal solution to the LP relaxation. However, again our feeling is that this might lead to less powerful fixing than is achievable in the dual context.

6   Preliminary Computational Results

We implemented a first version of an ABC algorithm as described in the previous sections. For the LP solution, we used the CPLEX 7.0 callable library of ILOG.

We tested the algorithm on the same set of 50 *multi-dimensional 0-1 knapsack* instances we already considered in [17] and [18]. Specifically, the problems are of the form $\max\{c^T x : Ax \leq b, x \in \{0,1\}^n\}$, where $c \in Z_+^n$, $A \in Z_+^{m \times n}$ and $b \in Z_+^m$, which were randomly generated as follows. For any pair (n, m) with $n \in \{5, 10, 15, 20, 25\}$ and $m \in \{5, 10\}$, we constructed 5 random instances whose objective function coefficients are integers generated uniformly between 1 and 10. Moreover, for the instances with $m = 5$, the left-hand side coefficients are also integers generated uniformly between 1 and 10, while for the instances with $m = 10$, the left-hand side coefficients have a 50% chance of being an integer generated uniformly between 1 and 10, but also have a 50% chance of being zero. That is, these instances are sparse. In all cases the right hand side of each constraint was set to half the sum of the left hand side coefficients[1].

Cutting. For the cutting part of the algorithm we used the same policy developed in [17]: we generate primally violated *lifted cover* inequalities, heuristically separated as described in [17] since their separation has been shown to be NP-hard (see, [16] for details), and when we are not able to find any of them we resort to generating a round of Gomory fractional cuts strengthened as in Letchford & Lodi [18]. After 25 consecutive rounds of Gomory fractional cuts, in order to avoid numerical problems, we branch.

Branching. The branching tree is explored in a *depth-first* search. Differently from what we described in Section 4 we do not complement the current $\bar{x}$, thus we distinguish between a *left-branch*, which is fixing a variable to the same value assumed in the incumbent solution, and a *right-branch* which corresponds to implicitly imposing the other value through the addition of a set of inequalities as described in Section 4.

There are two interesting things to point out. First, the choice of exploring the tree in depth-first manner implies that just two sets must be maintained during the search: we call N_{left} (resp. N_{right}) the set of the variables which are fixed according to (resp. at the contrary of) their value in the incumbent solution. These sets correspond to sets N_0 and N_1 of (1), and an augmentation is simply handled by moving the variables which are currently contained in N_{right} to N_{left} (and, obviously, by manipulating some of the constraints added in the right-branches). Second, as alluded to in Section 4, only in the case of the *first* right-branch a set of constraints must be added, and specifically $n - |N_{left}| - 1$ constraints. In the following right-branches, it is enough to change an **inequality** (previously introduced in the first right-branch) into an **equality**. A straightforward example of this behavior is the following.

Example. Assume that the incumbent solution is such that $\bar{x}_i = 1$ and $\bar{x}_j = 0$. Since the first right-branch, at node h, has been performed on variable x_i, a constraint $x_j + x_i \leq 1$ has been added at node h. If at node k (for which node h is a father), we want to explore the right-branch associated to variable x_j, it is enough to transform the previously added constraint into $x_j + x_i = 1$.

---

[1] This is well-known to lead to non-trivial instances of the multi-dimensional 0-1 knapsack problem.

Other implementation details and further advances will be discussed in following studies. Concerning the results on the multi-dimensional knapsack instances, they are reported in Table 1.

Table 1. ABC algorithm. Preliminary results on multi-dimensional knapsack instances.

m	n	Aug.	nodes to Opt.	nodes	primal LCIs	overall cuts
	5	2.0	1.0	1.0	4.4	5.2
	10	4.8	1.0	1.0	7.8	16.8
5	15	8.6	26.4	48.6	25.2	736.0
	20	10.0	2.4	189.4	42.6	2503.2
	25	13.8	682.0	765.0	69.0	2196.6
	5	0.8	0.6	1.0	5.0	5.0
	10	4.6	1.0	1.0	9.0	10.0
10	15	7.0	1.8	8.6	13.0	172.6
	20	10.0	8.6	97.4	34.0	669.6
	25	13.2	597.8	1521.0	77.0	8815.0

Table 1 reports for each pair (m, n), the average results on 5 instances of the number of augmentation performed by the algorithm (Aug.) starting by the trivial solution with all the variables set to 0, the number of branching nodes to find the optimal solution (nodes to Opt.), and the number of branching nodes to prove optimality (nodes). The last two columns of the table refer to cuts by reporting the average number of cuts added (overall cuts) and the average number of primal lifted cover inequalities separated (primal LCIs).

With respect to the algorithm outlined in Section 3.4, we resort to generating rounds of primally-violated Gomory fractional cuts as soon as primal separation fails, and not only when an integer infeasible point is encountered. Each round of Gomory cuts contains at most 25 cuts tight to $\bar{x}$. Moreover, step 1. of the algorithm above is disregarded, in the sense that we start with the trivial 0-solution, and we do not apply during the search any heuristic to improve the current solution.

The results obtained show that an augment-and-branch-and-cut algorithm is a viable way of solving 0-1 ILPs (and MILPs) provided that all the sophisticated techniques developed for standard branch-and-cut algorithms were also implemented in this context. Indeed, a comparison with a general-purpose branch-and-cut framework like CPLEX 7.0 is totally unfair at the moment due to the much larger arsenal of cuts and to the great level of software engineering development of the current version of CPLEX. Just to give an idea, however, by avoiding in CPLEX presolve, primal heuristic, and cut generation but including cover and Gomory fractional inequalities and performing a depth-first search, the average number of nodes required for the 5 instances with $m = 5$ and $n = 25$ is 80.2 with respect to 765.0 reported in Table 1, i.e., almost ten times fewer.

Finally, we also preliminary tested our ABC implementation on three of the very famous instances proposed by Crowder, Johnson & Padberg [6], namely p0033, p0040 and p0201. For these instances we start by the first integer solution given by CPLEX. The algorithm works quite well on the two smallest instances: it is able to prove optimality for the starting solution of p0040 without any branching and just 2 primal cuts, while the initial solution of p0033 is augmented twice and solved with 367 nodes. On p0201, instead, degeneracy becomes a severe problem. More than 95% of the time is spent performing degenerate pivots, and in this situation we resort to branching so that the number of nodes becomes huge. This suggests that the method could be improved by some form of anti-stalling device, or by periodically 'purging' the LP of unnecessary non-binding constraints. Further progress on this issue will be discussed in future studies.

7   Conclusion

We have examined how to perform separation and branching within the primal context and we have seen that, just as Weismantel suggested, it is possible to integrate augmentation, branching and cutting within a single framework, at least for (mixed) 0-1 problems. We have also shown that most of the components of a sophisticated branch-and-cut algorithm have a primal counterpart. Moreover, we have implemented and computationally tested the first version of an ABC algorithm which is a completely new approach to integer programming. The effectiveness of such an approach clearly needs to be proved on harder instances and future (actually, current) work will be devoted to obtaining sophisticated ABC algorithms and ad hoc implementations for specific classes of problems (e.g., for the TSP).

References

1. E. Balas, S. Ceria & G. Cornuéjols (1993) A lift-and-project cutting plane algorithm for mixed 0-1 programs. *Math. Program.* 58, 295–324.
2. E. Balas, S. Ceria & G. Cornuéjols (1996) Mixed 0-1 programming by lift-and-project in a branch-and-cut framework. *Mgt. Sci.* 42, 1229–1246.
3. E. Balas, S. Ceria, G. Cornuéjols & N. Natraj (1996) Gomory cuts revisited. *Oper. Res. Lett.* 19, 1–9.
4. A. Ben-Israel & A. Charnes (1962) On some problems of diophantine programming. *Cahiers du Centre d'Études de Recherche Opérationelle* 4, 215–280.
5. A. Caprara & M. Fischetti (1997) Branch-and-Cut Algorithms. In M. Dell'Amico, F. Maffioli & S. Martello (eds.) *Annotated Bibliographies in Combinatorial Optimization*, pp. 45-64. New York, Wiley.
6. H. Crowder, E.L. Johnson & M.W. Padberg (1983) Solving large-scale zero-one linear programming problems. *Oper. Res.* 31, 803–834.
7. F. Eisenbrand, G. Rinaldi & P. Ventura (2001) 0/1 primal separation and 0/1 optimization are equivalent. *Working paper*, IASI, Rome.
8. R.T. Firla, U.-U. Haus, M. Köppe, B. Spille & R. Weismantel (2001) Integer pivoting revisited. *Working paper*, Institute of Mathematical Optimization, University of Magdeburg.

9. R.S. Garfinkel & G.L. Nemhauser (1972) *Integer Programming*. New York: Wiley.
10. F. Glover (1968) A new foundation for a simplified primal integer programming algorithm. *Oper. Res.* 16, 727–740.
11. R.E. Gomory (1958) Outline of an algorithm for integer solutions to linear programs. *Bulletin of the AMS* 64, 275–278.
12. R.E. Gomory (1960) An algorithm for the mixed-integer problem. *Report RM-2597*, Rand Corporation, 1960 (Never published).
13. M. Grötschel, L. Lovász & A.J. Schrijver (1988) *Geometric Algorithms and Combinatorial Optimization*. Wiley: New York.
14. U.-U. Haus, M. Köppe & R. Weismantel (2000) The integral basis method for integer programming. *Math. Meth. of Oper. Res.* 53, 353–361.
15. A.H. Land & A.G. Doig (1960) An automatic method for solving discrete programming problems. *Econometrica* 28, 497–520.
16. A.N. Letchford & A. Lodi (2001) Primal separation algorithms. *Technical Report* OR/01/5. DEIS, University of Bologna.
17. A.N. Letchford & A. Lodi (2002) Primal cutting plane algorithms revisited. *Math. Methods of Oper. Res.*, to appear.
18. A.N. Letchford & A. Lodi (2002) Strengthening Chvátal-Gomory Cuts and Gomory fractional cuts. *Oper. Res. Letters*, to appear.
19. G.L. Nemhauser & L.A. Wolsey (1988) *Integer and Combinatorial Optimization*. New York: Wiley.
20. M.W. Padberg & M. Grötschel (1985) Polyhedral computations. In E. Lawler, J. Lenstra, A. Rinnooy Kan, D. Shmoys (eds.). *The Traveling Salesman Problem*, John Wiley & Sons, Chichester, 307–360.
21. M.W. Padberg & S. Hong (1980) On the symmetric travelling salesman problem: a computational study. *Math. Program. Study* 12, 78–107.
22. M.W. Padberg & G. Rinaldi (1991) A branch-and-cut algorithm for the resolution of large-scale symmetric travelling salesman problems. *SIAM Rev.* 33, 60–100.
23. A. Schulz, R. Weismantel & G. Ziegler (1995) 0-1 integer programming: optimization and augmentation are equivalent. In: *Lecture Notes in Computer Science*, vol. 979. Springer.
24. R. Thomas (1995) A geometric Buchberger algorithm for integer programming. *Math. Oper. Res.* 20, 864–884.
25. R. Urbaniak, R. Weismantel & G. Ziegler (1997) A variant of Buchberger's algorithm for integer programming. *SIAM J. on Discr. Math.* 1, 96–108.
26. R. Weismantel (1999) Private communication.
27. R.D. Young (1965) A primal (all-integer) integer programming algorithm. *J. of Res. of the National Bureau of Standards* 69B, 213–250.
28. R.D. Young (1968) A simplified primal (all-integer) integer programming algorithm. *Oper. Res.* 16, 750–782.

A Procedure of Facet Composition for the Symmetric Traveling Salesman Polytope

Jean François Maurras and Viet Hung Nguyen*

LIM, Université de la Mediterranée, 163 Avenue de Luminy, 13288 Marseille, France

Abstract. We propose a new procedure of facet composition for the Symmetric Traveling Salesman Polytope(STSP). Applying this procedure to the well-known *comb* inequalities, we obtain completely or partially known classes of inequalities like *clique-tree*, *star*, *hyperstar*, *ladder* inequalities for STSP. This provides a proof that a large subset of hyperstar inequalities which are until now only known to be valid, are indeed facets defining inequalities of STSP and this also generalizes ladder inequalities to a larger class. Finally, we describe some new facet defining inequalities obtained by applying the procedure.

1 Introduction

The Symmetric Traveling Salesman Polytope $STSP^n$ is the convex hull of the incidence vectors of all the Hamiltonian cycles of a complete undirected graph with n nodes. This polytope is associated with the well-known *traveling salesman problem* which is one of the most basic NP-hard combinatorial optimization problems. Thus, a considerable amount of research work has been devoted to characterizing or describing classes of facet defining inequalities for $STSP^n$.

Due to the very complex structure of $STSP^n$ it is very difficult to describe all inequalities which define facets for this polytope. A technique used to simplify the description is to define some operations on the inequalities that allow the derivation of new inequalities from others that have already been characterized. One of these operations is the *composition* of inequalities that produces new facet defining inequalities by merging two or more inequalities, known to be facet defining, which satisfy some conditions. These inequalities are called *blocks* of the composition. Naddef and Rinaldi [9] described a procedure of facet composition for $STSP$ called *2-sum* composition. This procedure helped to derive a large class of facet defining inequalities called *regular parity path-tree*. Some others facet composition procedures have been given by Queyranne and Wang [13], [12]. Blocks inequalities of these procedures need only to satisfy some simple conditions that can be easily verified.

Let us recall some known facts about $STSP^n$. Let $K_n = (V_n, E_n)$, it is proved that the affine hull of the hamiltonian cycles of K_n are n degree equations corresponding to the n vertices of K_n. These equations denote the degree constraints which say that any hamiltonian cycle contains exactly two edges of $\omega(v)$, the

---

* Current address: LIP6, 4 place Jussieu 75005 Paris

M. Jünger et al. (Eds.): Combinatorial Optimization (Edmonds Festschrift), LNCS 2570, pp. 134–146, 2003.

set of incident edges with v, for all $v \in V_n$. Thus, the dimension d of $STSP^n$ is equal to $n(n-1)/2 - n$.

Given a valid inequality I of $STSP^n$, a hamiltonian cycle whose incidence vector satisfies I at equality is called a *tight* hamiltonian cycle with respect to I. A set of tight hamiltonian cycles with respect to I that contains d hamiltonian cycles whose incidence vectors are affinely independent, is called a *kernel* of I. We have the following:

> The inequality I defines a facet of $STSP^n$ if and only if there exists a kernel of I.

If I is a facet defining inequality, in general the total number of tight hamiltonian cycles with respect to I is significantly greater than d and there are many different kernels of I. Intuitively, kernels whose cardinality is close to d express better the specific structure of the facet defined by I than kernels whose cardinality is much bigger than d. More precisely, the hamiltonian cycles of a kernel whose cardinality close to d usually share more common properties than the hamiltonian cycles of a kernel with greater cardinality. In this paper, we develop this intuition into a procedure of facet composition for $STSP^n$. The procedure can be resumed as follows:

> We are given two or more facet defining inequalities and the corresponding kernels whose hamiltonian cycles have some identical property. Composing these inequalities is then composing the corresponding kernels to obtain a kernel of the new inequality. Thus, this new inequality defines a facet of $STSP$ provided that it is valid.

Thus our procedure aims to exploit the specific structure of inequalities to extend them.

To illustrate the procedure, we apply it to the well-known *comb* inequalities [2][5]. This allows us to obtain completely or partially known classes of inequalities like *clique-tree* [6], *star* [4], *hyperstar* [3], *ladder* [1] inequalities for STSP. The readers can find descriptions of these inequalities in [10] and [7]. In our knowledge, until now there is no proof of facet defining for hyperstar inequalities. By our procedure of composition, we provide a proof of being facet defining for a large subset of hyperstar inequalities. We also give a generalization of ladder inequalities and some other new facet defining inequalities.

The paper is organized as follows. First, we introduce some notations and notions. We then describe the procedure of composition. Finally we apply the procedure to *comb* inequalities. Because of space limit, we will only present an extended abstract, a more complete version can be found in [11]

2   Definitions and Notations

Let $G = (V, E)$ be a undirected graph. The edge between two vertices u and v in G will be denoted as uv. For $X \subset V$, $E(X)$ denotes the set of all edges $uv \in E$

such that both $u, v \in X$ and $\omega(X)$ denotes the set of all edges $uv \in E$ such that $u \in X$ and $v \in V \setminus X$. For $X, X' \subset V$ and $X \cap X' = \emptyset$, $(X : X')$ denotes the set of edges for which one endnode belongs to X and the other belongs to X'. For $Y \subset E$, $V(Y)$ denotes the set of all vertices u such that at least one edge of Y is incident with u. A cycle C of G will be considered as a set of edges. Let α be a vector of $\mathbb{R}^E$, for $Y \subset E$, let $\alpha(Y) = \sum_{e \in Y} \alpha_e$.

Let $K_n = (V_n, E_n)$ be a complete undirected graph with n vertices. Let $I \equiv \gamma^T \mathbf{x} \leq \gamma_0$ be a facet defining inequality of $STSP^n$.

Definition 1 (Subclique). *A subset $X \subset V_n$ is a subclique if $\gamma_e = \gamma_{e'}$ for all $e, e' \in E(X)$. This common coefficient is called γ_X.*

Definition 2 (Critical set). *A subset $D \subset E_n$ is a critical set with respect to I if there is a kernel of I such that every hamiltonian cycle in I contains at least one edges of D. Indeed, it is conceivable that for some edges of D, there is no hamiltonian cycle in I which contains them.*

Definition 3 (δ-critical set). *A subset $D \subset E_n$ is δ-critical set with respect to I if there is a kernel of I such that every hamiltonian cycle in I contains <u>exactly</u> δ mutually non-adjacent edges of D, i.e. a matching of cardinality δ of D.*

Definition 4 (Co-critical sets). *Let $D_1, D_2, \ldots, D_k$ be a collection of edge sets such that D_i is δ_i-critical set for all $i = 1, 2, \ldots, k$. These sets are co-critical if there is a kernel of I whose each member contains a matching of cardinality δ_i of D_i for all $i = 1, 2, \ldots, k$ and the union of these matchings form a matching of cardinality $\delta_1 + \delta_2 + \ldots + \delta_n$ of the set $\bigcup_{i=1}^{k} D_i$.*

Let $G = (V, E)$ be a undirected graph. Let $STSP(G)$ be the symmetric traveling salesman polytope defined on G. We assume that the dimension d_G of $STSP(G)$ is equal to $|E| - |V|$.

Definition 5 (3-cycle forest). *A 3-cycle forest $\mathcal{F} \subset E$ of G is a spanning subgraph of G whose connected components contain exactly one cycle of length 3.*

An inequality $\beta^T \mathbf{x} \leq \beta_0$ or an equality $\beta^T \mathbf{x} = \beta_0$ is said $\mathcal{F}$-canonic if $\beta_e = 0$ for all $e \in \mathcal{F}$ where $\mathcal{F}$ is a 3-cycle forest of G. We have the following lemma

Lemma 1. *Let $\mathcal{F}$ be 3-cycle forest of G and $I \equiv \gamma^T \mathbf{x} \leq \gamma_0$ be a valid inequality of $STSP(G)$. Let $\beta^T \mathbf{x} = \beta_0$ be a $\mathcal{F}$-canonic equality.*

(i) *If all the tight hamiltonian cycles with respect to I also satisfy the equality $\beta^T \mathbf{x} = \beta_0$ and fixing k coefficients $\beta_{e_1}, \beta_{e_2}, \ldots, \beta_{e_k}$ at 0 where $e_i \notin E \setminus \mathcal{F}$, implies $\beta = \mathbf{0}$ (the zero vector), then I defines a $(d_G - k)$-face of the $STSP(G)$.*

(ii) *Inversely, if I defines a $(d_G - k)$-face of $STSP(G)$ and there is $(d_G - k + 1)$ tight hamiltonian cycles affinely independent with respect to I which also satisfy the equality $\beta^T \mathbf{x} = \beta_0$ then by fixing k coefficients $\beta_{e_1}, \beta_{e_2}, \ldots, \beta_{e_k}$ at 0 where $e_i \notin E \setminus \mathcal{F}$, one can show that $\beta = \mathbf{0}$.*

The proof of this lemma is based on the fact that $\mathcal{F}$ is a column base of the incidence matrix of G. Therefore, there is a unique inequality (up to a positive multiple) $\mathcal{F}$-canonic that is equivalent with I.

3   Facet Composition by Mean of δ-Critical Sets

Let us consider two complete undirected graphs $K_{n_1} = (V_{n_1}, E_{n_1})$ and $K_{n_2} = (V_{n_2}, E_{n_2})$. Let $I_1 \equiv \gamma^T \mathbf{x} \leq \gamma_0$ be a facet defining inequality of $STSP^{n_1}$ and $I_2 \equiv \alpha^T \mathbf{x} \leq \alpha_0$ be a facet defining inequality of $STSP^{n_2}$.
Suppose that there are a subclique X_1 of K_{n_1} and a subclique X_2 of K_{n_2} such that:

H1 $|X_1| = |X_2| = 2\delta + 1$ and $\gamma_{X_1} = \gamma_{X_2} = \rho$.
H2 $E(X_1)$ is a δ-critical set with respect to I_1 and $E(X_2)$ is a δ-critical set with respect to I_2.
H3 Let $U_1 \subset E(X_1)$ (respectively $U_2 \subset E(X_1)$) be any set of $\delta + 1$ edges consisting of a matching of cardinality δ plus an edge such that $V(U_1) = X_1$ (respectively $V(U_2) = X_2$). There is a tight hamiltonian cycle of K_{n_1} (respectively K_{n_2}) with respect to I_1 (respectively I_2) that contains the edges of U_1 (respectively U_2).

By uniting X_1 and X_2 into a unique set X, we obtain a new graph $G = (V, E)$ from K_{n_1} and K_{n_2} such that $V = (V_{n_1} \setminus X_1) \cup (V_{n_2} \setminus X_2) \cup X$ and $E = (E_{n_1} \setminus E(X_1)) \cup (E_{n_2} \setminus E(X_2)) \cup E(X)$.
Set $d_1 = n_1(n_1 - 1)/2$ and $d_2 = n_2(n_2 - 1)/2$. We have $|E| = d_1 + d_2 - |E(X)|$.
Let $\eta \in \mathbb{R}^E$ such that:

 - $\eta_e = \gamma_e$ for all $e \in E_{n_1}$.
 - $\eta_e = \alpha_e$ for all $e \in E_{n_2}$.

Let us call $d_G = |E| - n_1 - n_2 + |X| = |E| - n_1 - n_2 + 2\delta + 1$ the dimension of $STSP(G)$.

Theorem 1. *The inequality $I \equiv \eta^T \mathbf{x} \leq \gamma_0 + \alpha_0 - (2\delta + 1)\rho$ defines a $(d_G - 2)$-face of $STSP(G)$.*

Proof. We shall show the validity of I and then explain briefly the outline of the proof for the inequality I to define a $(d_G - 2)$-face.
Let C be any hamiltonian cycle of G.
We inspect this cycle by going through it in a given direction D. We meet the vertices of G which belong to either $V_{n_1} \setminus X$ or $V_{n_2} \setminus X$ or X. In this inspection, as the vertices of X are like bridges between the vertices of $V_{n_1} \setminus X$ and the vertices of $V_{n_2} \setminus X$, so we will go through alternatively a maximal directed path contained in $E_{n_1} \setminus E(X)$ and a maximal directed path contained in E_{n_2}. Thus we obtain a set $\mathcal{P}$ of maximal directed disjoint paths which are contained either in $E_{n_1} \setminus E(X)$ or in E_{n_2}. The paths in $E_{n_1} \setminus E(X)$ begin and end with a vertex of $V_{n_1} \setminus X$. The paths in E_{n_2} begin and end with a vertex of X. The total number of these paths is obviously positive and even. Suppose that there are $2p$ such paths. Let $U \subset X = \{u_1, u_2, \dots, u_{2p}\}$, the set of the end vertices of all maximal directed paths in E_{n_2} of $\mathcal{P}$. For i odd and $1 \leq i \leq 2p - 1$, assume that following the direction D, between u_i and u_{i+1} there is a maximal directed path in E_{n_2}. We derive that for j even and $2 \leq j \leq 2p - 2$, there is a maximal directed path in $E_{n_1} \setminus E(X)$ between u_j and u_{j+1} and there is a maximal directed path in

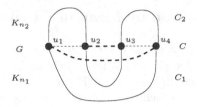

Fig. 1. A hamiltonian cycle C of the graph G and its partition into C_1 and C_2.

$E_{n_1} \setminus E(X)$ between u_{2p} and u_1. Replace the maximal directed paths in E_{n_2} by the edges u_1u_2, u_3u_4, $\ldots$, $u_{2p-1}u_{2p}$, we obtain a cycle C_1 (non necessarily hamiltonian) of K_{n_1}.

Similarly, replace the maximal directed paths in E_{n_1} by the edges u_2u_3, u_4u_5, $\ldots$, $u_{2p}u_1$, we obtain a cycle C_2 (non necessarily hamiltonian) of K_{n_2}.

Remark 1. The edges of $C_1 \cap E(U)$ (respectively $C_2 \cap E(U)$) form a matching of $E(U)$. The edge of $(C_1 \cup C_2) \cap E(U)$ form a hamiltonian cycle of the subgraph induced by U. In the case where $|U| = 2$, this cycle reduces to an edge counted two times (a loop if we give a direction to the two edges). The cycle C_1 is called a complement of C_2 with respect to U and vice-versa.

Thus, all hamiltonian cycles of G are obtained from a pair of complementary cycles C_1 and C_2.

We calculate now the value of $\eta(C)$. We have obviously $\eta(C) = \eta(C_1) + \eta(C_2) - 2p\rho$.

Let us call W_1 (respectively W_2) the set of vertices of C_1 (respectively C_2) which belong to X. The sets W_1, W_2 and U form a partition of X. Thus, we have

$$|W_1| + |W_2| = 2\delta + 1 - 2p.$$

Note that we can complete the cycle C_1 to a hamiltonian cycle C_1' of K_{n_1} by replacing a particular edge in $C_1 \cap E(U)$, for example u_1u_2, by a path with all the vertices of W_2 as the interior vertices and u_1, u_2 as the two ends. We obtain:

$$\eta(C_1') = \eta(C_1) + |W_2|\rho.$$

Similarly for the cycle C_2, we obtain a hamiltonian cycle C_2' of K_{n_2} and

$$\eta(C_2') = \eta(C_2) + |W_2|\rho.$$

Since these two cycles are respectively hamiltonian cycles of K_{n_1} and K_{n_2}, we also have:

$$\eta(C_1') = \gamma(C_1') \leq \gamma_0 \quad \text{and} \quad \eta(C_2') = \alpha(C_2') \leq \alpha_0.$$

All above equations allow us to derive that $\eta(C) \leq \gamma_0 + \alpha_0 - (2\delta + 1)\rho$. Hence, inequality I is valid.

We characterize now tight hamiltonian cycles of G with respect to I.

Let $\mathcal{K}_1$ be a set of tight hamiltonian cycles of K_{n_1} with respect to I_1 which is a kernel corresponding to the critical set $E(X_1) = E(X)$. Similarly, let $\mathcal{K}_2$ be a set of tight hamiltonian cycles of K_{n_2} with respect to I_2 which is a kernel corresponding to the critical set $E(X_2) = E(X)$.

Let $U = \{u_1, u_2, \dots, u_{2\delta}\}$ be a subset of cardinality 2δ of X and let $C_1' \in \mathcal{K}_1$. The δ non-adjacent edges of C_1' in $E(X)$ are denoted by $u_i u_{i+1}$ for all $i = 1, \dots, 2\delta-1$ and odd. Let w be the only vertex of X that does not belong to U and $C_2' \in \mathcal{K}_{n_2}$ that contains the edges $u_2 w$, $w u_4$ and $u_{2\delta} u_1$ and the $\delta - 2$ non-adjacent edges $u_j u_{j+1}$ for all $j = 4, \dots, 2\delta - 2$ and even. Replacing the edges $u_2 w$ and $w u_4$ by the edge $u_2 u_4$ in C_2', we obtain a $(n_2 - 1)$-cycle C_2 of K_{n_2}. Note that C_2 is the complement of C_1' with respect to U and the hamiltonian cycle C of G obtained from C_1' and C_2, is tight with respect to I since

$$\eta(C) = \eta(C_1') + \eta(C_2') - 2\delta\rho = \gamma_0 - (\alpha_0 - \rho) - 2\delta\rho = \gamma_0 - \alpha_0 - (2\delta + 1)\rho.$$

The cycle C_2 is called a maximal complement of C_1'.

Symmetrically, we can obtain a tight hamiltonian cycle of G from a tight hamiltonian cycle in $\mathcal{K}_{n_2}$ and one of its maximal complement.

Let $\mathcal{K}$ be the set of all tight hamiltonian cycles with respect to I that are built like above. We hope that $\mathcal{K}$ is a kernel of I. But in fact, we will show that $\mathcal{K}$ is nearly a kernel of I, i.e. $\mathcal{K}$ contains $d_G - 1$ tight hamiltonian cycles affinely independent. We give the outline of the proof. Let $X = \{x_1, x_2, \dots, x_{2\delta+1}\}$ and $\mathcal{F}$ be a 3-cycle forest of G such that the path $x_1 x_2 \dots x_i x_{i+1} \dots x_{2\delta+1}$ and the edge $x_1 x_3$ form a connected component of $\mathcal{F}$. Let $\beta^T \mathbf{x} = \beta_0$ be a $\mathcal{F}$-canonic equality. By definition, we have $\beta_e = 0$ for all $e \in \mathcal{F}$.

Suppose that all tight hamiltonian cycles of G with respect to I also satisfy $\beta^T \mathbf{x} = \beta_0$. By using the tight hamiltonian cycles in $\mathcal{K}$, we show that $\beta_e = 0$ for all $e \in E(X)$. We also show that the tight hamiltonian cycles of K_{n_1} in $\mathcal{K}_1$ satisfy the equality $\sum_{e \in E_{n_1}} \beta_e \mathbf{x}_e = \beta_1$, and the tight hamiltonian cycles of K_{n_2} in $\mathcal{K}_2$ satisfy the equality $\sum_{e \in E_{n_2}} \beta_e \mathbf{x}_e = \beta_2$, where $\beta_1 + \beta_2 = \beta_0$.

Since $\mathcal{F} \cap E_{n_1}$ is a 3-cycle forest of K_{n_1} and $\mathcal{K}_{n_1}$ is a kernel of I_1, thus according to the second part of Lemma 1, by fixing a coefficient β_{e_1} ($e_1 \in E_{n_1} \setminus \mathcal{F}$) at 0, we can derive that $\beta_e = 0$ for all $e \in E_{n_1}$.

The same holds for a coefficient β_{e_2} ($e_2 \in E_{n_2} \setminus \mathcal{F}$) and β_e for all $e \in E_{n_2}$.

Thus, fixing two coefficients β_{e_1} and β_{e_2} at 0 implies that $\beta_e = 0$ for all $e \in E$. According to the first part of Lemma 1, we conclude that I defines a (d_G-2)-face of $STSP(G)$.

We discuss now how to transform this $(d_G - 2)$-face into a facet, i.e. a $(d_G - 1)$-face of $STSP(G)$. Let us call *crossing edges*, the edges in $(V_{n_1} \setminus X : V_{n_2} \setminus X)$ which do not belong to E. We give several sufficient conditions on the facet defining inequalities I_1 and I_2 such that we can add a crossing edge e to the graph G, give a value to the coefficient η_e and the new inequality $I \equiv \eta^T \mathbf{x} \leq \eta_0$ defined now on the new graph G is valid and there are two tight hamiltonian cycles affinely independent containing the edge e. Then it is easy to see that these two hamiltonian cycles are affinely independent to the hamiltonian cycles in $\mathcal{K}$. Thus I defines a facet of $STSP(G)$.

Suppose that there exists in G two vertices $v_1 \in V_{n_1} \setminus X$ and $v_2 \in V_{n_2} \setminus X$ such that

- For all $i = 2, 3, \ldots, 2\delta + 1$, we have $\eta_{v_1 x_1} > \eta_{v_1 x_i}$ and $\eta_{v_2 x_1} > \eta_{v_2 x_i}$. For all $i, j = 2, 3, \ldots, 2\delta + 1$ and $i \neq j$, $\eta_{v_1 x_i} = \eta_{v_1 x_j}$ and $\eta_{v_2 x_i} = \eta_{v_2 x_j}$. In addition, $\eta_{v_1 x_1} + \eta_{v_2 x_i} = \eta_{v_2 x_1} + \eta_{v_1 x_i}$.
- For an edge $v_1 x_i$ (resp. $v_2 x_i$) where $i = 2, 3, \ldots, 2\delta + 1$
  - any tight hamiltonian cycle with respect to I_1 (resp. I_2) containing $v_1 x_i$ (resp. v_2) contains a path between v_1 (resp. v_2) and x_1 which does not contain other vertices de X but x_1.
  - any not tight hamiltonian cycle C_1 (resp. C_2) with respect to I_1 (resp. I_2) containing $v_1 x_i$ (resp. v_2), $\eta(C_1) = \gamma_0 - \eta_{v_1 x_1}$ (resp. $\eta(C_2) = \alpha_0 - \eta_{v_2 x_1}$).
- For any δ non-adjacent edges in $E(X)$, there exist tight hamiltonian cycles with respect to I_1 (resp. I_2) of K_{n_1} (resp. K_{n_2}) which contain these edges and the edge $v_1 x_1$ (resp. $v_2 x_1$).

Proposition 1 *Add the edge $v_1 v_2$ to G and set*

$$\eta_{v_1 v_2} := \eta_{v_1 x_1} + \eta_{v_2 x_i} - \rho (= \eta_{v_2 x_1} + \eta_{v_1 x_i} - \rho) \text{ where } x_i \in X \setminus \{x_1\}$$

The new inequality $I \equiv \eta^T \mathbf{x} \leq \eta_0$ defines a facet for $STSP(G)$.

Proof. Because of space limit, we omit the proof of the validity of I.

We will specify now two tight hamiltonian cycles affinely independent containing the edge $v_1 v_2$. Let $X_1 = X \cup \{v_1\}$ and $X_2 = X \cup \{v_2\}$, we can see that these sets are of even cardinality. Let C_1 be a tight hamiltonian cycle with respect to I_1 containing an edge $v_1 x_j$ where $j = 2 \ldots 2\delta + 1$. By definition, C_1 contains a path between v_1 and x_1 and thus we can find a cycle C_2 which is complementary of C_1 with respect to X_1 such that the cycle C_2^* obtained by replacing in C_2 the edges $v_1 v_2$ and $v_1 x_1$ by the edge $v_2 x_1$, is a tight hamiltonian cycle with respect to I_2. Let $M = C_1 \cup C_2 \cap E(X_1 \setminus v_1)$, we have $V(M) = X$ and $|M| = 2\delta$. We can see that the edge set $C_1 \cup C_2^* \cup v_1 v_2 \setminus M \setminus \{v_2 x_1, v_1 x_j\}$ forms a hamiltonian cycle C containing $v_1 v_2$ in G. We have

$$\eta(C) = \eta(C_1) + \eta(C_2) - \eta_{v_1 v_2} - \eta(M) - \eta_{v_2 x_1} - \eta_{v_1 x_j} =$$

$$= \alpha_0 + \gamma_0 + \eta_{v_2 x_1} + \eta_{v_1 x_j} - \rho - 2\delta\rho - \eta_{v_2 x_1} - \eta_{v_1 x_j} = \alpha_0 + \gamma_0 - (2\delta + 1)\rho.$$

Thus C is tight with respect to I. Symmetrically, we can derive from C_2' a tight hamiltonian cycle with respect to I_2 containing $v_2 x_j$ ($j = 2 \ldots 2\delta + 1$) and from C_1' a tight hamiltonian cycle with respect to I_1 containing $v_1 x_1$ which is complementary of C_2' with respect to X_2, a tight hamiltonian cycle C' in G with respect to I. It is easy to see that C and C' are affinely independent.

Consider the complete graph K_n weighted by a vector γ.

Definition 6 (Perfect subclique). *A subclique $X \subset V_n$ is a perfect subclique if for all $v \in V_n \setminus X$, all components of γ corresponding to the edges in $(v : X)$ are equal.*

We can generalize Theorem 1 by replacing vertices in X_1 and X_2 by perfect subcliques with respect to I.

Definition 7 (Super-set). *Let $X = \{S_1, \ldots, S_{2\delta+1}\}$ where the sets $S_i \in X$ are disjoint perfect cliques. Let*

$$E_X = \bigcup_{i,j=1}^{2\delta+1} (S_i : S_j).$$

The set X is super-set if all components of γ corresponding to the edges in E_X are equal. Let us call γ_{E_X} this common coefficient.

Definition 8 (Super-matching). *An edge set $D \subset E_n$ is a super matching if $D \subset E_X$ and for all $i = 1, \ldots, 2\delta + 1$ $|D \cap \omega(S_i)| \leq 1$.*

Definition 9 (Super δ-critical set). *The set E_X is a super δ-critical set with respect to a facet-defining I of $STSP^n$ if there exists a kernel of I whose each member contains a super-matching of cardinality δ of X.*

Definition 10 (Super co-critical set). *Let $D_1, \ldots, D_k$ be a collection of edge sets such that for all $i = 1, \ldots, k$, D_i is super δ_i-critical with respect to I. These sets are super co-critical if there exists a kernel of I such that every hamiltonian cycle in I contains a super-matching of cardinality δ_i of D_i for all $i = 1, \ldots, k$ and the union of these super-matchings form a super-matching of cardinality $\delta_1 + \ldots + \delta_k$ of the set $\cup_{i=1}^{k} D_i$.*

Definition 11 (k-vertex-critical). *[9] A vertex $u \in V_n$ is k-vertex-critical if for all hamiltonian cycles of maximum weight with respect to γ, we have $\gamma(C) = \gamma_0 - k$.*

Let $I_1 \equiv \gamma^T \mathbf{x} \leq \gamma_0$ and $I_2 \equiv \alpha^T \mathbf{x} \leq \alpha_0$ be respectively a facet-defining inequality of $STSP^{n_1}$ and $STSP^{n_2}$.

Suppose that there exists a super-set $X_1 = \{S_1, \ldots, S_{2\delta+1}\}$ in K_{n_1} and a super-set $X_2 = \{T_1, \ldots, T_{2\delta+1}\}$ in K_{n_2} such that

(i)   For all $i = 1, \ldots, 2\delta + 1$, we have $|S_i| = |T_i|$ and $\gamma_{S_i} = \alpha_{T_i}$. Then we set $\rho_i = \gamma_{S_i} = \alpha_{T_i}$ for all $u \in S_i$, u is ρ_i-vertex-critical with respect to I_1 and for all $v \in T_i$, v is ρ_i-vertex-critical with respect to I_2.

(ii) E_{X_1} is super δ-critical with respect to I_1 and E_{X_2} is super δ-critical with respect to I_2. In addition $\gamma_{E_{X_1}} = \alpha_{E_{X_2}} = \rho$.

(iii) For $U_1 \subset E_{X_1}$ such that $|U_1| = \delta + 1$ and for all $i = 1, \ldots 2\delta + 1$ we have $U_1 \cap S_i \neq \emptyset$, there is a tight hamiltonian cycle with respect to I_1 which contains U_1. Similarly, for $U_2 \subset E_{X_2}$ such that $|U_2| = \delta + 1$ and for all $i = 1, \ldots 2\delta + 1$ we have $U_2 \cap T_i \neq \emptyset$, there is a tight hamiltonian cycle with respect to I_2 which contains U_2.

By uniting X_1 and X_2 into a unique set X (this is done by uniting successively S_i and T_i for all $i = 1, \ldots, 2\delta + 1$ into a unique set R_i), we obtain the graph G. We have $X = \{R_1, \ldots, R_2\}$ where $R_i = (S_i \equiv T_i)$ and

$$|E(G)| = d_1 + d_2 - \sum_{i=1}^{2\delta+1} |E(S_i)| - |E_X|.$$

Let $\eta \in \mathbb{R}^{E(G)}$ such that

- $\eta_e := \gamma_e \ \forall e \in E_{n_1}$,
- $\eta_e := \alpha_e \ \forall e \in E_{n_2}$.

Let d_G be the dimension of $STSP(G)$, we can see that $d_G = |E(G)| - n_1 - n_2 + \sum_{i=1}^{2\delta+1} |R_i|$.

Theorem 2. *The inequality*

$$I \equiv \eta^T \mathbf{x} \leq \gamma_0 + \alpha_0 - (2\delta + 1)\rho - \sum_{i=1}^{2\delta+1} (|S_i| - 1)\rho_i,$$

defines a $(d_G - 2)$-face of $STSP(G)$.

The sufficient condition so that I defines a facet of $STSP(G)$ becomes much simpler than the case of Proposition 1 : there must be at least one subclique of more than one vertex in X.

Proposition 1. *If there exists <u>at least</u> a set R_i such that $|R_i| \geq 2$ and $\rho_i \neq \rho$ then the inequality I defines a facet of $STSP(G)$.*

4   Applications to Comb Inequalities

Let us consider a complete undirected graph $K_n = (V_n, E_n)$. Let H, T_1, T_2, $\ldots, T_{2m+1}$ be subsets of V_n such that

- $m \geq 1$,
- $T_i \cap T_j \neq \emptyset$ with $1 \leq i \neq j \leq 2m + 1$,
- $T_i \cap H \neq \emptyset$ and $T_i \setminus H \neq \emptyset$ with $1 \leq i \leq 2m + 1$.

The inequality

$$I \equiv \mathbf{x}(E(H)) + \sum_{i=1}^{2m+1} \mathbf{x}(E(T_i)) \leq |H| + \sum_{i=1}^{2m+1} |T_i| - m - 1$$

defines a comb inequality of $STSP^n$.

The set H is called *handle* and the sets T_i are called *teeth*. If $|T_i \cap H| = 1$ for all $i = 1, 2, \ldots, 2m + 1$, these inequalities are called *Chvátal comb* since they have been introduced by Chvátal [2]. The general comb inequalities have been studied by Grötschel and Padberg [5].

Let $T = \bigcup_{j=1}^{2m+1} T_i$ and $\bar{H} = V_n \setminus H$. For each tooth T_i, let $Z_i = T_i \cap H$ and $\bar{Z}_i = T_i \cap \bar{H}$.

Let $U = \{Z_1, Z_2, \ldots Z_{2m+1}\}$ and $\bar{U} = \{\bar{Z}_1, \bar{Z}_2, \ldots, \bar{Z}_{2m+1}\}$.

Let R be the family of sets which are composed by the sets Z_i and all the singletons $\{y\}$ where $y \in H \setminus T$. Let $\bar{R}$ be the family of sets which are composed by the sets $\bar{Z}_i$ and all the singletons $\{\bar{y}\}$ where $\bar{y} \in H \setminus T$. For a comb inequality I, we will consider two following types of super-sets :

- Those whose elements belong to R. Let $\mathcal{P}$ be a collection of these sets.
- Those whose elements belong to $\bar{R}$. Let $\mathcal{Q}$ be a collection of these sets.

Suppose that $\mathcal{P} = \{P_1, P_2, \ldots, P_p\}$ and $\mathcal{Q} = \{Q_1, Q_2, \ldots, Q_q\}$ and the following conditions are satisfied by the super-sets in $\mathcal{P}$ and $\mathcal{Q}$:

(i)   For all $i = 1, \ldots, p$, $|P_i| = 2p_i + 1$ where $p_i \leq 1$. For all $j = 1, \ldots, q$, $|Q_i| = 2q_j + 1$ where $q_j \leq 1$.
(ii) For all $i = 1, \ldots, p$, $P_i \cap U \neq \emptyset$ and
    - if $P_i \subsetneq U$ then for all P_j with $j \neq i$, $P_i \cap P_j = \emptyset$.
    - otherwise, i.e. either $P_i = U$ or $P_i \backslash U \neq \emptyset$, then for all $j \neq i$, $|P_j \cap P_i| \leq 1$.
    Similarly, we have the same conditions for $\bar{U}$ and the super-sets $Q_j \in Q$.
(iii) If $p \geq 2$ and $q \geq 2$, for all sets $P_i \in \mathcal{P}$ and $Q_j \in \mathcal{Q}$ such that $|P_i| = |Q_j|$, $P_i \subsetneq U$ and $Q_j \subsetneq \bar{U}$, the number of pair $Z_k, \bar{Z}_k$ such that $Z_k \in P_k$ and $\bar{Z}_k \in Q_j$ is less than or equal to $|P_i| - 2$.

Theorem 3. *The subsets E_{P_i} and E_{Q_j} are respectively super p_i-critical and super q_j-critical with respect to I for all $i = 1, 2, \ldots, p$ and $j = 1, 2, \ldots, q$. In addition, these sets are all super co-critical.*

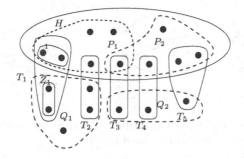

Fig. 2. An example of collections $\mathcal{P}$ and $\mathcal{Q}$ for a comb inequality.

Now we can apply the procedure of facet composition to comb inequalities.

By uniting two super critical sets of $\bar{R}$ of two comb inequalities, we can obtain a clique tree inequality.

We describe briefly combinatorial structure of star, hyperstar and ladder inequalities. Star inequalities are like comb inequalities but the handle H becomes

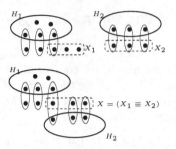

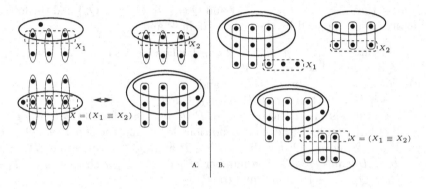

Fig. 3. A clique tree inequality obtained by composing two comb inequalities

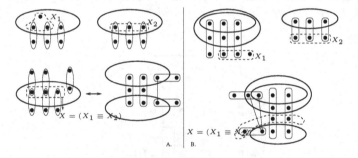

Fig. 4. A. A star inequality obtained by composing two comb inequalities. B.A hyperstar inequality obtained by composing a star inequality and a comb inequality

Fig. 5. A. A ladder inequality obtained by composing two comb inequalities. B. A generalized ladder inequality having nested handles.

a nested set. An application of our method to comb inequalities (by uniting the handles) gives a nice subset of star inequalities called *multiple-handled comb*. These inequalities are special cases of *path* inequalities that have been proved facet-inducing for $STSP^n$ by Naddef and Rinaldi [8]. Hyperstar inequalities accept multiple handles like clique-tree inequalities but handles can be nested. An application of our methods to comb and multi-handled comb inequalities (by uniting (subsets) of handles or (subsets) of teeth) gives hyperstar inequalities

whose coefficients of edges in handles are 1 and in these inequalities a tooth intersects a handle also intersects every handle containing it (we call *non-degenerate*
tooth such a tooth, the teeth that do not have this property will be call *degenerate* teeth) . Figure 4 illustrates an example of the application.

Ladder inequalities defined in [1] have only two handles H_1 and H_2 with some
teeth intersecting both of them and two others T_1 and T_2 intersecting respectively H_1 and H_2. There are *pending* edges between $T_1 \cap H_1$ and $T_2 \cap H_2$. Among
teeth intersecting both handles, there are also *degenerate* teeth which have no
vertices outside the handles. An application of our method to comb inequalities
gives ladder inequalities having no degenerate teeth and generalize them to have
more than two handles. An example is illustrated in Figure 5.

Consider a tooth T_i of a comb inequality I such that $|T_i \setminus H| \geq 2$. Let P be
a maximal super-set of odd cardinality which is composed by the set Z_i and
singletons of $T_i \setminus H$. Let $|P| = 2\delta + 1$, we have the following theorem

Theorem 4. *E_P is a super δ-critical set with respect to I.*

We can apply our procedure of composition to these super critical sets and by
this operation, we obtain new facet-defining inequalities of $STSP^n$ which allow
an even number of teeth intersecting a handle. Figure 6 gives an example of such
inequality.

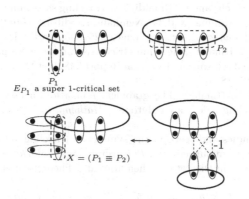

Fig. 6. A new facet-inequality of $STSP^n$.

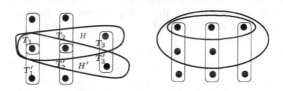

Fig. 7. A composition of two comb inequalities giving a star inequality with a degenerate tooth.

5 Remark

To generate star, hyperstar inequalities and ladder inequalities having degenerate teeth, we need to extend the composition method which allows to unit one vertex belonging to a handle and one vertex outside the handles. An example is given in Figure 7.

References

1. S. Boyd, W. Cunningham, M. Queyranne, and Y. Wang. Ladders for travelling salesmen. *SIAM Journal on Optimization*, 5:408–420, 1995.
2. V. Chvátal. Edmonds polytopes and weakly hamiltonian graphs. *Mathematical Programming*, 5:29–40, 1973.
3. B. Fleischman. Cutting planes for the symmetric traveling salesman problem. Technical report, Universitat Hamburg, 1987.
4. B. Fleischman. A new class of cutting planes for the symmetric travelling salesman problem. *Mathematical Programming*, 40:225–246, 1988.
5. M. Grötschel and M. Padberg. On the symmetric traveling salesman problem I: inequalities. *Mathematical Programming*, 16:265–280, 1979.
6. M. Grötschel and W. Pulleyblank. Clique tree inequalities and the symmetric traveling salesman problem. *Mathematics of operations research*, 11:537–569, 1986.
7. M. Jünger, G. Reinelt, and G. Rinaldi. The traveling salesman problem. In M. Ball, T. Magnanti, C. Monma, and G. Nemhauser, editors, *Handbook on Operations Research and Management Science*, pages 225–330. North Holland, 1995.
8. D. Naddef and G. Rinaldi. The symmetric traveling salesman polytope: new facets from the graphical relaxation. Technical Report 248, Instituto di Analisi dei Sistemi ed Informatica, 1988.
9. D. Naddef and G. Rinaldi. The graphical relaxation: a new framework for the symmetric traveling salesman polytope. *Mathematical Programming*, 58:53–88, 1993.
10. D. Naddef. Handles and teeth in the symmetric traveling salesman polytope. In W. Cook and P. Seymour, editors, *Polyhedral Combinatorics*, volume 1, pages 61–74. DIMACS series in Discrete Mathematics and Theoretical of Computer Science, AMS-ACM, 1990.
11. V. H. Nguyen. *Polyèdres de cycles : Description, Composition et Lifting de Facettes*. PhD thesis, Université de la Méditerranée, Marseille, 2000.
12. M. Queyranne and Y. Wang. Facet-tree composition for symmetric travelling salesman polytopes. Technical Report 90-MSC-001, Faculty of Commerce and Business Administration, University of British Columbia, 1990.
13. M. Queyranne and Y. Wang. Composing facets of symmetric travelling salesman polytopes. Technical report, Faculty of Commerce and Business Administration, University of British Columbia, 1991.

Constructing New Facets of the Consecutive Ones Polytope

Marcus Oswald and Gerhard Reinelt

Institut für Informatik, Universität Heidelberg, Im Neuenheimer Feld 368,
D-69120 Heidelberg, Germany,
{Marcus.Oswald,Gerhard.Reinelt}@Informatik.Uni-Heidelberg.De

Abstract. In this paper we relate the consecutive ones problem to the betweenness problem by pointing out connections between their associated polytopes. We will prove some results about the facet structure of the betweenness polytope and show how facets of this polytope can be used to generate facets of the consecutive ones polytope. Furthermore, the relations with the consecutive ones polytopes will enable us to conclude that the number of facets of the consecutive ones polytope only grows polynomially if the number of columns is fixed. This gives another proof of the fact that the consecutive ones problem is solvable in polynomial time in this case.

1 Introduction

A 0/1-matrix A has the *consecutive ones property for rows* if its columns can be ordered in such a way that in every row the ones occur consecutively. For convenience we just say that A *is C1P*. Whereas it is easy to check if a matrix is C1P, it is NP-hard to compute for a given 0/1-matrix the minimum number of entries to be switched for obtaining the consecutive ones property. This is the so-called *consecutive ones problem*. If there are individual penalties for switching an entry and if we want to minimize the total cost for converting the matrix to be C1P we speak about the *weighted consecutive ones problem* (WC1P).

The input of the *betweenness problem* consists of a set of n objects $1, 2, \ldots, n$ and a set $\mathcal{B}$ of betweenness conditions. Every element of $\mathcal{B}$ is a triple (i, j, k) requesting that object j should be placed between objects i and k. The task is to find an ordering of the objects such that as few betweenness conditions as possible are violated. If violations are penalized by weights, we call the problem of finding an ordering which minimizes the sum of penalties the *weighted betweenness problem* (WBWP). Note that there can be non-betweenness conditions as well requiring that a certain object should not be placed between two objects. These conditions can be dealt with easily, so we do not discuss them here.

Both the WC1P and a variant of the WBWP occur as models in computational biology. In [2] and [3] first branch-and-cut approaches for these two problems are presented.

We review some definitions for the consecutive ones problem. For a 0/1-matrix A with m rows and n columns let $\chi^A = (a_{11}, \ldots, a_{1n}, \ldots, a_{m1}, \ldots, a_{mn})$ be its characteristic vector. We define the *consecutive ones polytope* as

M. Jünger et al. (Eds.): Combinatorial Optimization (Edmonds Festschrift), LNCS 2570, pp. 147–157, 2003.
© Springer-Verlag Berlin Heidelberg 2003

$$P_{\text{C1}}^{m,n} = \text{conv}\{\chi^A \mid A \text{ is an } (m,n)\text{-matrix with C1P}\}.$$

It is easy to see that $P_{\text{C1}}^{m,n}$ has full dimension $m \cdot n$.

We do not want to discuss $P_{\text{C1}}^{m,n}$ in detail here, but mention only that trivial lifting is possible for $P_{\text{C1}}^{m,n}$ and that the trivial inqualities $x_{ij} \geq 0$ and $x_{ij} \leq 1$ are facet-defining for all polytopes. Proofs of these theorems and an integer programming formulation of WC1P that consists of facet-defining inequalities only are given in [3].

In section 2 we discuss some aspects of the betweenness polytope. In particular, we prove a trivial lifting theorem for this polytope. In section 3 we define a master polytope which will allow us to point out some relations between the two polytopes. We show how facets of the betweenness polytope induce facets of the consecutive ones polytope. Based on this observation we then show in section 4 that the consecutive ones polytope only has a polynomial number of facets if the number of columns is fixed.

2   The Betweenness Polytope

In the following we use indices $i(j)k$ (*betweenness triples*) for pairwise different objects i, j and k, indicating that we consider whether object j is between objects i and k or not. Since the indices $i(j)k$ and $k(j)i$ are equivalent, we only use $i(j)k$ such that $i < k$. In vectors, triples are ordered lexicographically, i.e., we use the order

$$1(2)3, 1(2)4, \ldots, n-1(n-2)n.$$

For each permutation π of $n \geq 3$ elements $1, \ldots, n$ and each betweenness triple $i(j)k$ we define an indicator $\chi_{i(j)k}^{\pi}$ which is 1 if and only if the element j lies between the elements i and k in the permutation π and 0 otherwise. The $3\binom{n}{3}$-dimensional characteristic *betweenness vector* associated with a permutation π is

$$\chi^{\pi} = (\chi_{1(2)3}^{\pi}, \chi_{1(2)4}^{\pi}, \cdots, \chi_{n-1(n-2)n}^{\pi}).$$

The *betweenness polytope* P_{BW}^n, $n \geq 3$, is the convex hull of all betweenness vectors, i.e.,

$$P_{\text{BW}}^n = \text{conv}\{\chi^{\pi} \mid \pi \text{ is a permutation of } \{1, \ldots, n\}\}.$$

It is easy to show that the following is true.

Lemma 1. *For an arbitrary point* $x = (x_{1(2)3}, \ldots, x_{n-1(n-2)n}) \in P_{\text{BW}}^n$ *and three pairwise different* i, j, k, $1 \leq i, j, k \leq n$, *the betweenness equation*

$$x_{i(j)k} + x_{i(k)j} + x_{j(i)k} = 1$$

holds.

This lemma characterizes exactly (up to linear combinations) all equations that are valid for P_{BW}^n.

Theorem 1. P_{BW}^n *has dimension* $2\binom{n}{3}$.

A proof of this theorem is given in [4].

Let $a^T x \leq a_0$ be valid for P_{BW}^n and $n' > n$. We say that the inequality $\bar{a}^T x \leq a_0$ for $P_{BW}^{n'}$ is obtained from $a^T x \leq a_0$ by *trivial lifting* if

$$\bar{a}_{i(j)k} = \begin{cases} a_{i(j)k} & \text{if } 1 \leq i, j, k \leq n, \\ 0 & \text{otherwise.} \end{cases}$$

Trivial lifting means that larger polytopes inherit all nontrivial facets of smaller polytopes. To prove this we need the following two lemmata.

Lemma 2. *Let* $a^T x \leq a_0$ *be facet-defining for* P_{BW}^n, $n \geq 4$. *For each pair* $i, j \in \{1, \dots, n\}$ *there is at least one triple* $e(f)g$ *with* $|\{i, j\} \cap \{e, f, g\}| \leq 1$ *and* $a_{e(f)g} \neq 0$.

Proof. Assume that there is a pair i, j such that for all triples $e(f)g$ with $|\{i, j\} \cap \{e, f, g\}| \leq 1$ we have $a_{e(f)g} = 0$. Then the inequality can be written as

$$\sum_{k \notin \{i,j\}} (a_{i(j)k} x_{i(j)k} + a_{i(k)j} x_{i(k)j} + a_{j(i)k} x_{j(i)k}) \leq a_0.$$

It is easy to find a permutation π such that for every k

$$a_{i(j)k} x_{i(j)k} + a_{i(k)j} x_{i(k)j} + a_{j(i)k} x_{j(i)k} = \beta_k,$$

where $\beta_k = \max\{a_{i(j)k}, a_{i(k)j}, a_{j(i)k}\}$. Therefore $a^T x \leq a_0$ is already implied by the $n - 2$ trivial inequalities $\beta_k(x_{i(j)k} + x_{i(k)j} + x_{j(i)k}) \leq \beta_k$ (which are in fact equations).

Lemma 3. *Let* $a^T x \leq a_0$ *be facet-defining for* P_{BW}^n, $n \geq 4$. *For each pair* $i, j \in \{1, \dots, n\}$ *there is a vector* x^* *with*

$$x^*_{e(f)g} = \begin{cases} \frac{1}{2} & \text{if } \{i, j\} = \{e, f\} \text{ or } \{i, j\} = \{f, g\}, \\ 0 & \text{if } \{i, j\} = \{e, g\}, \\ \star & \text{otherwise.} \end{cases}$$

and x^* *can be written as an affine combination of betweenness vectors* χ^{π_i} *which satisfy* $a^T \chi^{\pi_i} = a_0$.

Proof. Because of Lemma 2 there is at least one triple $e(f)g$ with $a_{e(f)g} \neq 0$ and $x^*_{e(f)g} = \star$. After setting all the other "$\star$"-entries in an arbitrary way only fulfilling all the betweenness equations we can always choose $x^*_{e(f)g}$, $x^*_{f(e)g}$ and $x^*_{e(g)f}$ in such a way that $x^*_{e(f)g} + x^*_{f(e)g} + x^*_{e(g)f} = 1$ and $a^T x^* = a_0$ holds. Note that these entries not necessarily have to lie between 0 and 1. Since $a^T x \leq a_0$ defines a facet of P_{BW}^n, there must be an affine combination of the vertices of the facet which represents x^*.

We are now ready to prove the lifting theorem.

Theorem 2. *Let $a^T x \leq a_0$ be facet-defining for P_{BW}^n, $n \geq 4$ and let $n' > n$. If $a^T x \leq a_0$ is trivially lifted then the resulting inequality defines a facet of $P_{BW}^{n'}$.*

Proof. It is sufficient to show the theorem for $n' = n + 1$. Let $S(n)$ denote the set of permutations of $\{1, 2, \ldots, n\}$.

Since all permutations $\pi' \in S(n+1)$ contain a permutation $\pi \in S(n)$, the inequality $a^T x \leq a_0$ remains valid for P_{BW}^{n+1}. We only have to show that all equations $b^T x = b_0$ that hold for all $x \in P_{BW}^{n+1}$ which satisfy $a^T x \leq a_0$ with equality are multiples of $a^T x = a_0$. In our case it is even sufficient to show that $b_{e(f)g} = b_{f(e)g} = b_{e(g)f}$ for all triples e, f, g with $n + 1 \in \{e, f, g\}$, because in this case we can reduce the coefficients to 0 by adding suitable multiples of betweenness equations. Afterwards the equation can be trivially downlifted. And since $a^T x \leq a_0$ is facet-defining for P_{BW}^n, $b^T x = b_0$ must be a multiple of $a^T x = a_0$.

We proceed as follows. We show that for each pair $i, j \in \{1, \ldots, n\}$ the equation $b_{i(j)n+1} = b_{i(n+1)j} = b_{j(i)n+1}$ holds. We construct the vector x^* according to lemma 3. There are l permutations $\pi_1, \ldots, \pi_l \in S(n)$ satisfying $a^T \chi^{\pi_h} = a_0$ and

$$x^* = \sum_{h=1}^{l} d_h \chi^{\pi_i} \text{ with } \sum_{h=1}^{l} d_h = 1.$$

W.l.o.g. we can choose the permutations in such a way that the elements i occur before j (otherwise we can convert the order without changing the betweenness vectors). Now we construct l permutations $\pi'_{1_a}, \ldots, \pi'_{l_a} \in S(n+1)$ from the permutations $\pi_1, \ldots, \pi_l$ by inserting the element $n + 1$ directly before i and another l permutations $\pi'_{1_b}, \ldots, \pi'_{l_b} \in S(n+1)$ by inserting the element $n + 1$ directly after j. So we have

$$\pi'_{h_a} = (\ldots, n+1, i, \ldots, j, \ldots) \text{ and } \pi'_{h_b} = (\ldots, i, \ldots, j, n+1, \ldots)$$

Due to this construction, $b^T \chi^{\pi_{h_a}} = b^T \chi^{\pi_{h_b}} = b_0$ holds for the betweenness vectors of all of these permutations. Summing up all the equations and inserting the values of x^*, we obtain after some calculations that

$$0 = \sum_{h=1}^{l} (d_h b^T \chi^{\pi_{h_a}} - d_h b^T \chi^{\pi_{h_b}})$$

$$= \sum_{h=1}^{l} d_h b^T (\chi^{\pi_{h_a}} - \chi^{\pi_{h_b}})$$

$$= \ldots$$

$$= b_{n+1(i)j} - b_{i(j)n+1}.$$

Take for example the contribution of the triples $i(k)n+1$, $k \neq j$ to this sum. Since $\chi^{\pi_{h_a}}_{i(k)n+1} = 0$ and $\chi^{\pi_{h_b}}_{i(k)n+1} = \chi^{\pi_{h_b}}_{i(k)j}$ we get

$$\sum_{h=1}^{l} \sum_{k \neq j} (d_h b_{i(k)n+1} \chi^{\pi_{h_a}}_{i(k)n+1} - d_h b_{i(k)n+1} \chi^{\pi_{h_b}}_{i(k)n+1})$$

$$= -\sum_{k \neq j} b_{i(k)n+1} \sum_{h=1}^{l} d_h \chi^{\pi_{h_b}}_{i(k)j}$$

$$= -\sum_{k \neq j} b_{i(k)n+1} x^*_{i(k)j}$$

$$= -\sum_{k \neq j} b_{i(k)n+1} \cdot 0$$

$$= 0$$

The calculations of the other triples work similarly.

To get the second relation we construct permutations π'_{h_c} and π'_{h_d} where

$$\pi'_{h_c} = (\ldots, i, n+1, \ldots, j, \ldots) \text{ and } \pi'_{h_d} = (\ldots, i, \ldots, n+1, j, \ldots).$$

Here we obtain

$$0 = \sum_{h=1}^{l} d_h b^T (\chi^{\pi_{h_a}} - \chi^{\pi_{h_c}} - \chi^{\pi_{h_d}} + \chi^{\pi_{h_b}})$$

$$= \ldots$$

$$= b_{n+1(i)j} - 2 b_{i(n+1)j} + b_{i(j)n+1}.$$

From both relations we get the desired result $b_{i(j)n+1} = b_{i(n+1)j} = b_{j(i)n+1}$.

In contrast to the consecutive ones polytope, trivial inequalities only define facets in the smallest case. In [4] it is shown that the complete linear description of P^3_{BW} is given by the betweenness equation $x_{1(2)3} + x_{1(3)2} + x_{2(3)1} = 1$ and the three trivial inequalities $x_{1(2)3} \geq 0$, $x_{1(3)2} \geq 0$ and $x_{2(1)3} \geq 0$, and that for $n \geq 4$ none of the trivial inequalities $x_{i(j)k} \geq 0$ or $x_{i(j)k} \leq 1$ is facet-defining.

Since we have equations, the same facet-defining inequality can be stated in various ways. Therefore we define a *normal form* with the property that two facet-defining inequalities define the same facet if and only if their normal forms coincide.

Definition 1. *A facet-defining inequality is in normal form if it has the following properties.*

i) The inequality is written as $\sum a_{i(j)k} x_{i(j)k} \geq a_0$.

ii) All coefficients $a_{i(j)k}$ are nonnegative coprime integers.

iii) At least one of the three coefficients $a_{i(j)k}$, $a_{i(k)j}$ and $a_{j(i)k}$ is zero for pairwise different elements i, j and k, $1 \leq i, j, k \leq n$.

It is easily seen that the normal form of a facet-defining inequality is unique and that it is easy to convert an inequality to normal form.

3   A Common Polytope

Both the feasible solutions of the WBWP and of the WC1P are based on the permutations of n elements. For examining relations between the two problems we define a master problem combining their constraints. Here we seek for a permutation satisfying betweenness conditions as well as having an associated matrix where the ones appear consecutively.

We say that a permutation π of n elements *establishes* the consecutive ones conditions of a 0/1-matrix $A \in \{0,1\}^{m \times n}$ (π establishes C1C of A) if in the permuted matrix $A' \in \{0,1\}^{m \times n}$ with $a'_{ri} = a_{r\pi^{-1}(i)}$ the ones occur consecutively in every row. The common betweenness and consecutive ones polytope $P_{\mathrm{BWC1}}^{m,n}$ is defined as

$$P_{\mathrm{BWC1}}^{m,n} = \mathrm{conv}\{(\chi^\pi, \chi^A) \mid A \in \{0,1\}^{m \times n} \text{ and } \pi \text{ establishes C1C of } A\}.$$

The single polytopes can be simply obtained from $P_{\mathrm{BWC1}}^{m,n}$. Namely, the projection of $P_{\mathrm{BWC1}}^{m,n}$ on the betweenness variables $x_{i(j)k}$ is the betweenness polytope P_{BW}^n and the projection on the consecutive ones variables x_{ri} is the consecutive ones polytope $P_{\mathrm{C1}}^{m,n}$. Of course, all valid inequalities for $P_{\mathrm{C1}}^{m,n}$ or P_{BW}^n remain valid for $P_{\mathrm{BWC1}}^{m,n}$. One can easily construct valid inequalities for $P_{\mathrm{BWC1}}^{m,n}$ that are formulated both on betweenness and on consecutive ones variables.

Let $A = (a_{ri})$ be an (m, n)-matrix, $n \geq 3$, with C1P. Then for all rows r of A, all betweenness triples $i(j)k$ of columns i, j, k of A and all permutations π that establish C1C of A we have

$$\chi_{i(j)k}^\pi \leq 2 - a_{r\pi^{-1}(i)} + a_{r\pi^{-1}(j)} - a_{r\pi^{-1}(k)}.$$

Based on this observation we can define so-called *linking constraints*

$$x_{i(j)k} \leq 2 - x_{ri} + x_{rj} - x_{rk}$$

which are valid for $P_{\mathrm{BWC1}}^{m,n}$ for all $r \in \{1, \dots, m\}$ and all betweenness triples $i(j)k$.

Intuitively there is a close relationship between the consecutive ones and the betweenness problem. We now establish a connection between the facets of the two polytopes by making use of the linking constraints. These constraints are used to eliminate betweenness variables and replace them by consecutive ones variables.

Since in the consecutive ones problem we actually deal with matrices we will denote a valid inequality for $P_{\mathrm{C1}}^{m,n}$ by $B \circ x \leq b_0$, where B is the coefficient matrix, x is the matrix of the variables and $B \circ x = \sum_{i=1}^m \sum_{j=1}^n b_{ij} x_{ij}$. Further we will use $B_{i\cdot}^T x_{i\cdot}$ instead of $\sum_{j=1}^n b_{ij} x_{ij}$.

Theorem 3. *Let $a^T x \geq a_0$ be a facet-defining inequality for P_{BW}^n, $n \geq 4$, in normal form. Further let m be the number of nonzero coefficients of a. We assign pairwise different numbers $r_{i(j)k} \in \{1, \dots, m\}$ to the betweenness triples $i(j)k$ with $a_{i(j)k} > 0$. Let the inequality $B \circ x \leq b_0$ be obtained by summing up $-a^T x \leq -a_0$ and all (scaled) linking constraints $a_{i(j)k} x_{i(j)k} \leq a_{i(j)k}(2 - x_{r_{i(j)k}i} + x_{r_{i(j)k}j} - x_{r_{i(j)k}k})$. Then $B \circ x \leq b_0$ is facet-defining for $P_{\mathrm{C1}}^{m,n}$.*

Proof. We have to show that all equations $C \circ x = c_0$ that hold for all $x \in P^{m,n}_{C1}$ which satisfy $B \circ x \leq b_0$ with equality are multiples of $B \circ x = b_0$. Since $B \circ x \leq b_0$ is a conical sum of $-a^T x \leq -a_0$ and some linking constraints, all these inequalities must be fulfilled with equality for all C1P-matrices A that satisfy $B \circ \chi^A = b_0$ and associated permutations π that establish C1C of A with betweenness vector χ^π. Now we compute the entries of the row $r_{i(j)k}$ of C. Since $a^T x \geq a_0$ is a facet of $P^n_{\mathrm{BW}}, n \geq 4$, there must be at least one vector (χ^A, χ^π) with $\chi^\pi_{i(j)k} = 1$ (otherwise all vertices of the facet would fulfill $x_{i(j)k} = 0$). Because equality must hold for the linking constraint $x_{i(j)k} \leq 2 - x_{r_{i(j)k}i} + x_{r_{i(j)k}j} - x_{r_{i(j)k}k}$, there are three possible combinations

$$(\chi^A_{r_{i(j)k}i}, \chi^A_{r_{i(j)k}j}, \chi^A_{r_{i(j)k}k}) \in \{(1,0,0), (0,0,1), (1,1,1)\}$$

This means row $r_{i(j)k}$ of χ^A written in the order of π can look like

$$\chi^A_{r_{i(j)k}\cdot} = \begin{cases} (1,\dots,1,1,1,0,0,\dots,0,0,0,\dots,0,0,0,0,0,\dots,0) \\ \quad\vdots \\ (0,\dots,0,1,1,0,0,\dots,0,0,0,\dots,0,0,0,0,0,\dots,0) \\ (0,\dots,0,0,1,0,0,\dots,0,0,0,\dots,0,0,0,0,0,\dots,0) \\ (0,\dots,0,0,1,1,0,\dots,0,0,0,\dots,0,0,0,0,0,\dots,0) \\ \quad\vdots \\ (0,\dots,0,0,1,1,1,\dots,1,0,0,\dots,0,0,0,0,0,\dots,0) \\ (0,\dots,0,0,0,0,0,\dots,0,0,1,\dots,1,1,1,0,0,\dots,0) \\ \quad\vdots \\ (0,\dots,0,0,0,0,0,\dots,0,0,0,\dots,0,1,1,0,0,\dots,0) \\ (0,\dots,0,0,0,0,0,\dots,0,0,0,\dots,0,0,1,0,0,\dots,0) \\ (0,\dots,0,0,0,0,0,\dots,0,0,0,\dots,0,0,1,1,0,\dots,0) \\ \quad\vdots \\ (0,\dots,0,0,0,0,0,\dots,0,0,0,\dots,0,0,1,1,1,\dots,1) \\ (0,\dots,0,0,1,1,1,\dots,1,1,1,\dots,1,1,1,0,0,\dots,0) \end{cases}$$

By substituting differences of suitable rows and the remainders of the corresponding matrices into the equation $C \circ x = c_0$ one easily can conclude that $C_{r_{i(j)k}l} = 0$ for $l \notin \{i, j, k\}$ and $C_{r_{i(j)k}i} = -C_{r_{i(j)k}j} = C_{r_{i(j)k}k} =: c_{i(j)k}$. This holds for all triples $i(j)k$ with $a_{i(j)k} > 0$. And since all linking constraints must be satisfied with equality we can construct the equation $\sum_{i(j)k} c_{i(j)k} x_{i(j)k} =: c^T x = \tilde{c}_0$ which is satisfied by all considered betweenness-vectors χ^π. But $a^T x \geq a_0$ is facet-defining for P^n_{BW} and therefore $c^T x = \tilde{c}_0$ must be a multiple of $a^T x = a_0$ which also means that $C \circ x = c_0$ is a multiple of $B \circ x = b_0$.

Consider for example the inequality

$$x_{1(2)3} + x_{1(3)2} + x_{2(1)4} + x_{3(1)4} \geq 1$$

which is facet-defining for P^4_{BW} and in normal form. The four linking constraints

$$-x_{1(2)3} \geq -2 + x_{11} - x_{12} + x_{13}$$
$$-x_{1(3)2} \geq -2 + x_{21} - x_{23} + x_{22}$$
$$-x_{2(1)4} \geq -2 + x_{32} - x_{31} + x_{34}$$
$$-x_{3(1)4} \geq -2 + x_{43} - x_{41} + x_{44}$$

are valid for $P^{4,4}_{\mathrm{BWC1}}$.

Summing up these five inequalities (thus eliminating the betweenness variables) and multiplying by -1 yields the inequality

$$\begin{pmatrix} 1 & -1 & 1 & 0 \\ 1 & 1 & -1 & 0 \\ -1 & 1 & 0 & 1 \\ -1 & 0 & 1 & 1 \end{pmatrix} \circ x \leq 7.$$

As already shown in [3]) this inequality is facet-defining for $P^{4,4}_{\mathrm{C1}}$.

This relation between facets of the two polytopes can be used for a new separation procedure for the WC1P. Assume we are given an LP-solution $x^* = (x^*_{ij})$ of the WC1P. First we compute a virtual LP-solution $y^* = (y^*_{i(j)k})$ for the WBWP by setting

$$y^*_{i(j)k} = \min_r \{2 - x^*_{ri} + x^*_{rj} - x^*_{rk}\}.$$

Now we can use any separation procedure for the WBWP to find betweenness-facets that violate y^*. For any facet we find in this way one can construct a facet for the WC1P which violates x^*.

4   The Consecutive Ones Problem for a Fixed Number of Columns

If we start with a facet $a^T x \geq a_0$ of P^n_{BW} in normal form, clearly the support of the constructed facet of $P^{m,n}_{\mathrm{C1}}$ has at most n columns. But what about the number of rows m? According to the construction, m is the number of nonzero coefficients of a. Since at least one of three coefficients is zero, we have $m \leq 2\binom{n}{3}$. Taking into account that these facets can be trivially lifted to facets of $P^{m',n}_{\mathrm{C1}}$ with $m' > m$ the total number of constructed facets of $P^{m,n}_{\mathrm{C1}}$ for fixed n and arbitrary m is $O(m^{2\binom{n}{3}})$ and thus polynomial in m.

Unfortunately, not every facet can be constructed in this way. Now we want to generalize this method to show the surprising result that for fixed n the total number of facets of $P^{m,n}_{\mathrm{C1}}$ grows polynomially in m with the consequence that WC1P is polynomially solvable for fixed n.

Lemma 4. *Let $a^T x \geq a_0$ be a facet-defining inequality for P_{BW}^n in normal form and $B \circ x \leq b_0$ a derived facet of $P_{\mathrm{C1}}^{m,n}$. Further let $Q = \{\pi \mid a^T \chi^\pi = a_0\}$ and $Q' = \{\pi \mid$ there exists $C \in \{0,1\}^{m \times n}$ with $B \circ C = b_0$ and π establishes C1C of $A\}$ be the sets of permutations which fulfill the respective inequalities with equality. Then $Q = Q'$ and for every $\pi \in Q$, every 0/1-matrix C with the property that π establishes C1C of C and every betweenness triple $i(j)k$ with nonzero coefficient $a_{i(j)k}$ and corresponding row $r = r_{i(j)k}$ the condition*

$$B_{r.}^T C_{r.} = a_{i(j)k}(2 - \chi_{i(j)k}^\pi)$$

holds.

Proof. According to the construction $B_{r.}$ has only 3 nonzero entries $b_{ri} = b_{rk} = a_{i(j)k}$ and $b_{rj} = -a_{i(j)k}$. It follows that $B_{r.}^T C_{r.} = a_{i(j)k}(c_{ri} - c_{rj} + c_{rk})$ and thus the condition is equivalent to a linking constraint. And all these linking constraints have to be fulfilled with equality since $B \circ x \leq b_0$ as conical combination of $-a^T x \leq -a_0$ and the linking constraints is fulfilled with equality.

A consequence of this observation is the fact that for any C1P matrix C which satisfies a facet-defining inequality $B \circ x \leq b_0$ with equality the values of $B_{r.}^T C_{r.}$ only depend on the row r of B and on a permutation π that establishes C1C of C but not on the remainder of the matrix B and not on the matrix C itself. A generalization of this observation is used to prove the following result.

Theorem 4. *The number of facets of $P_{\mathrm{C1}}^{m,n}$ for fixed n is $O(m^{n!/2})$.*

Proof. Let $B \circ x \leq b_0$ be any nontrivial facet-defining inequality for $P_{\mathrm{C1}}^{m,n}$. Since the zero matrix and all matrices consisting of zeroes except for one single entry are feasible solutions $b_0 > 0$ follows.

Our goal is to show that the support of B has at most $n!/2$ rows.

Let π be an arbitrary permutation with the property that there exists a matrix $C \in \{0,1\}^{m \times n}$ with $B \circ C = b_0$ and π establishes C1C of C. For each of these permutations π and every row r of B we define

$$m_r^\pi(B) = \max\{B_{r.}^T v \mid v \in \{0,1\}^{1 \times n} \text{ and } \pi \text{ establishes C1C of } v\}$$

as the maximum possible contribution of row r to the left hand side of the facet.

Note that if we denote by π' the permutation obtained when reading π in reverse order (i.e., $\pi'(i) = n + 1 - \pi(i)$, for $1 \leq i \leq n$), then $m_r^\pi(B) = m_r^{\pi'}(B)$ holds since the consecutive ones conditions are not affected by reversing the order.

Furthermore, for all C1P matrices C with $B \circ C = b_0$ and establishing permutation π the relation

$$B_{r.}^T C_{r.} = m_r^\pi(B)$$

holds for every row r. $B_{r.}^T C_{r.} \leq m_r^\pi(B)$ is clear from the definition of $m_r^\pi(B)$. If we assume that $B_{r.}^T C_{r.} < m_r^\pi(B)$ then we can construct a new C1P matrix C' by replacing the row r of C by the maximum row v from the above definition. But then $B \circ C' > b_0$ contradicting the validity of $B \circ x \leq b_0$.

Now let s be the number of rows of B and t be the number of considered permutations in an arbitrary order $\pi_1, \ldots, \pi_t$. We create the $s \times t$-matrix

$$M(B) = (m_{ij}(B)) = (m_i^{\pi_j}(B)).$$

Since $t \leq n!$ and since for every π_j there is a $\pi_{j'}$ with $m_{ij}(B) = m_{ij'}(B)$ for all rows i, the rank of $M(B)$ is at most $n!/2$ independently of the number s of rows.

Now assume that a facet-defining inequality $B \circ x \leq b_0$ is given with the number of rows s of B greater than $n!/2$. Because of rank $M(B) \leq n!/2$ at least one row r of $M(B)$ can be written as linear combination $M(B)_r. = \sum_{i \neq r} d_i M(B)_i.$ And since $B_r^T. C_r. = m_{rj}(B)$ holds for any C1P matrix C with $B \circ C = b_0$ and an establishing permutation π_j we get

$$B_r^T. C_r. = \sum_{i \neq r} d_i B_i^T. C_i..$$

This equation holds for every vertex C of $\{x \in P_{C1}^{m,n} \mid B \circ x = b_0\}$. And since it contains no constant coefficient it cannot be obtained by scaling the equation $B \circ x = b_0$ with $b_0 > 0$. Therefore the inequality $B \circ x \leq b_0$ cannot be facet-defining.

Thus the support of every facet-defining inequality for $P_{C1}^{m,n}$ has at most $n!/2$ rows and therefore each of these facets can be obtained by trivial lifting from a facet of $P_{C1}^{n!/2,n}$. Since the number of facets of $P_{C1}^{n!/2,n}$ is constant in m and the number of lifting possibilities for one facet is at most $\binom{m}{n!/2}$ the total number of facet-defining inequalities for $P_{C1}^{m,n}$ is $O(m^{n!/2})$.

As consequence we obtain that WC1P is solvable in polynomial time for fixed n.

Corollary 1. *WC1P is solvable in polynomial time for fixed n.*

Proof. According to Theorem 4 all facets of $P_{C1}^{m,n}$ can be obtained by trivial lifting from facets of $P_{C1}^{n!/2,n}$. Calculating these facets takes constant time in m. And for each of these facets there are at most $\binom{m}{n!/2}$ possibilities for trivial lifting. Thus we need time $O(m^{n!/2})$ to create a complete listing of all facets of $P_{C1}^{m,n}$ and therefore the WC1P is solvable in polynomial time for fixed n.

However, with a fairly simple algorithm we can even solve the problem in linear time (in m).

Namely, let the WC1P be formulated as $\max\{B \circ x \mid x \in \{0,1\}^{m \times n}$ is C1P$\}$. Now for each column permutation π and each row r of B we calculate $m_r^\pi(B)$. One calculation can be done in $O(n)$ with a scan line algorithm. Since

$$\max\{B \circ x \mid x \in \{0,1\}^{m \times n} \text{ is C1P}\}$$
$$= \max\{\sum_r m_r^\pi(B) \mid \pi \text{ is permutation of } \{1, \ldots, n\}\}$$

holds we are done. The total running time of this algorithm is $O(n!mn)$ and thus linear in m.

There are some remaining conjectures and open questions on the relations between the two polytopes.

We believe that the linking constraints are facet-defining for $P_{\mathrm{BWC1}}^{m,n}$. We verified this for $m = 1$ and $3 \leq n \leq 5$ but could not yet prove the general case.

A further interesting question is if there is a reverse construction from C1P-facets to BWP-facets. However the inequality

$$\begin{pmatrix} 1 & 1 & -1 & -1 \\ 1 & -1 & 1 & -1 \\ 1 & -1 & -1 & 1 \end{pmatrix} \circ x \leq 5$$

is facet-defining for $P_{\mathrm{C1}}^{3,4}$ but cannot be derived from a facet of P_{BW}^{4} by our construction.

The constructive proof of Theorem 4 leads to the idea of generalizing the betweenness variables $\chi_{i(j)k}^{\pi}$ by defining new variables

$$\chi_{w}^{\pi} = \max\{w^{T}v \mid v \in \{0,1\}^{n} \text{ and } \pi \text{ establishes C1C of } v\}$$

for suitable vectors $w \in \mathbf{Z}^{n}$. For $n = 4$ we found 10 variables (8 betweenness variables and 2 additional ones) which turn out to be a good choice. At least all facets of $P_{\mathrm{C1}}^{m,4}$ we have investigated so far can be derived in this way.

Moreover it is interesting to study separation for the betweenness polytope because separation routines for this polytope can be employed for the consecutive ones polytope as well.

References

1. T. Christof, A. Loebel (1998) PORTA - A Polyhedron Representation Algorithm. www.informatik.uni-heidelberg.de/groups/comopt/software/PORTA
2. T. Christof, M. Oswald and G. Reinelt (1998) Consecutive Ones and A Betweenness Problem in Computational Biology. *Proceedings of the 6th IPCO Conference, Houston*, 213–228
3. M. Oswald and G. Reinelt (2000) Polyhedral Aspects of the Consecutive Ones Problem, *Proceedings of the 5th Conference on Computing and Combinatorics, Sydney*, 373–382
4. M. Oswald and G. Reinelt (2001) Some Relations Between Consecutive Ones and Betweenness Polytopes, to appear in: *Proceedings of OR2001, Selected Papers, Duisburg*

A Simplex-Based Algorithm for 0-1 Mixed Integer Programming*

Jean-Philippe P. Richard[1], Ismael R. de Farias[2], and George L. Nemhauser[1]

[1] School of Industrial and Systems Engineering, Georgia Institute of Technology,
Atlanta GA 30332-0205, USA
[2] Center for Operations Research and Econometrics, 34 Voie du Roman Pays, 1348
Louvain-La-Neuve, Belgium

Abstract. We present a finitely convergent cutting plane algorithm for
0-1 mixed integer programming. The algorithm is a hybrid between a
strong cutting plane and a Gomory-type algorithm that generates violated facet-defining inequalities of a relaxation of the simplex tableau
and uses them as cuts for the original problem. We show that the cuts
can be computed in polynomial time and can be embedded in a finitely
convergent algorithm.

1 Introduction

Gomory [7] in the 1950's pioneered the idea of using cutting planes to solve integer programs. In his approach, valid inequalities are generated from rows of the
currently fractional simplex tableau. An advantage of this method is that a violated inequality can always be found quickly and the resulting algorithm can be
proven to converge in a finite number of iterations. Although appealing concep-
tually, this algorithm is not effective in practice. Branch-and-cut schemes that
use strong cutting planes have proven to be much more effective in solving 0-1
mixed integer programs. They proceed by relaxing the constraint set into a polytope whose structure has been studied and for which families of facet-defining
inequalities (facets for short) are known. A separation procedure, usually heuristic, is called to generate a violated facet of the relaxed polytope, which is a cut
for the initial problem. This idea was used by Crowder, Johnson and Padberg
[5] with the theoretical foundation coming from the polyhedral studies of the
0-1 knapsack polytope by Balas [1], Balas and Zemel [3], Hammer, Johnson and
Peled [11] and Wolsey [20]. Subsequently there have been many applications of
this approach that imbeds strong cuts into a branch-and-bound algorithm, see
surveys by Johnson, Nemhauser and Savelsvergh [12] and Marchand, Martin,
Weismantel and Wolsey [13]. The main disadvantage of this approach is that
the separation procedure may not return any cut, at which point partial enumeration is required. In this paper we introduce an algorithm that is a hybrid
between these two approaches. We generate cuts from the simplex tableau and
therefore are able to produce a violated inequality at every step of the algorithm.

* This research was supported by NSF grants DMI-0100020 and DMI-0121495

M. Jünger et al. (Eds.): Combinatorial Optimization (Edmonds Festschrift), LNCS 2570, pp. 158–170, 2003.
© Springer-Verlag Berlin Heidelberg 2003

Moreover, our cuts are always facets for relaxations of the polytopes defined by the rows of the tableau and yield a finite simplex algorithm for 0-1 mixed integer programming. The cuts are generally less dense than Gomory cuts and can be computed in polynomial time, although Gomory cuts are less expensive to obtain. Our cuts are derived in Richard, de Farias and Nemhauser [18,19] and in order to understand their validity and facetial properties, it is necessary to refer to these papers.

We now introduce our basic relaxation. Let $M = \{1, \ldots, m\}$ and $N = \{1, \ldots, n\}$. Given the sets of positive integers $\{a_1, \ldots, a_m\}$ and $\{b_1, \ldots, b_n\}$ together with the positive integer d, let

$$S = \{(x, y) \in \{0, 1\}^m \times [0, 1]^n \mid \sum_{j \in M} a_j x_j + \sum_{j \in N} b_j y_j \leq d\}.$$

The *mixed 0-1 knapsack polytope* is $PS = conv(S)$. Note that, since m and n are positive integers, the knaspack inequality contains both continuous and integer variables. As long as the continuous variables are bounded, it is not restrictive to choose the bounds to be 0 and 1. Also, it is not restrictive to require the coefficients a_j and b_j to be positive since we can always complement variables. Finally we may assume that the coefficients a_j and b_j are smaller than d, otherwise x_j can be fixed to 0 and y_j can be rescaled. Given these assumptions, we have

Theorem 1. *PS is full-dimensional.*   □

Although we are not aware of any previous study of the mixed 0-1 knapsack polytope other than our current work in [18,19], valid and facet-defining inequalities for related polytopes have been known for quite some time. For example there are the mixed integer cuts introduced by Gomory [8], the MIR inequalities introduced by Nemhauser and Wolsey [16] and the mixed disjunctive cuts introduced by Balas, Ceria and Cornuéjols [2]. More closely related to our study is the "0-1 knapsack polytope with a single continuous variable" introduced by Marchand and Wolsey [14]. There are significant differences between the polyhedron of Marchand and Wolsey and PS, which are described in [18].

In Section 2 we propose a conceptual framework to generate cuts for 0-1 mixed integer problems, using some knowledge of PS. This technique requires an extensive use of lifting, including the lifting of continuous variables as studied in [18,19]. In Section 3 we present a family of facets for the mixed 0-1 knapsack polytope that can be used as tableau cuts for general 0-1 mixed integer problems and can be obtained in polynomial time. In Section 4 we sketch how these cuts are used in a finitely convergent algorithm.

2   Generating Cuts from the Simplex Tableau

In this section, we present a formal framework to obtain cuts from a relaxation of the simplex tableau. This procedure requires extensive use of lifting techniques and motivates the introduction of the following notation.

Let M_0, M_1 be two disjoint subsets of M, and N_0, N_1 be two disjoint subsets of N. Define

$$S(M_0, M_1, N_0, N_1) = \{ (x, y) \in S \mid x_j = 0 \, \forall j \in M_0, x_j = 1 \, \forall j \in M_1,$$
$$y_j = 0 \, \forall j \in N_0, \, y_j = 1 \, \forall j \in N_1 \}$$

and $PS(M_0, M_1, N_0, N_1) = conv(S(M_0, M_1, N_0, N_1))$.

Consider the general 0-1 mixed-integer program

$$\max \sum_{j \in M} c_j x_j + \sum_{j \in N} c_{m+j} y_j \tag{1}$$

$$s.t. \sum_{j \in M} a_{ij} x_j + \sum_{j \in N} b_{ij} y_j \le d_i \qquad \forall i \in H \tag{2}$$

$$x_j \in \{0, 1\} \qquad \forall j \in M \tag{3}$$

$$y_j \in [0, 1] \qquad \forall j \in N \tag{4}$$

where $H = \{1, \ldots, h\}$. Note that each row of (2) together with (3) and (4) has the form of S. We let Q be the set of solutions to (2)-(4) and $PQ = conv(Q)$.

Because we work with simplex tableaux, we introduce slacks and assume they are continuous. Since lower and upper bounds on the slack variables are easily obtained, they can be rescaled so that their domain is the interval $[0, 1]$ or substituted out if the bounds are equal. We can therefore replace (2) by

$$\sum_{j \in M} a_{ij} x_j + \sum_{j \in N} b_{ij} y_j + u_i y_{n+i} = d_i \qquad \forall i \in H \tag{5}$$

and (4) by

$$y_j \in [0, 1] \qquad \forall j \in \tilde{N} \tag{6}$$

where $\tilde{N} = N \cup \{n+1, \ldots, n+h\}$.

Now consider a solution to the LP relaxation of this problem. If none of the 0-1 variables is fractional, then the current solution is optimal. So we may assume that at least one of them is fractional. Note that nonbasic variables are either at their lower or upper bound, so that no integrality violations can occur for them. Therefore any 0-1 variable that is fractional in the current LP relaxation has to be basic. Assume the ith variable of the basis is 0-1 and fractional. The ith row of the simplex tableau can be written as

$$x_{B(i)} + \sum_{j \in M_0 \cup M_1} \tilde{a}_{ij} x_j + \sum_{j \in N_0 \cup N_1} \tilde{b}_{ij} y_j = f_i + \sum_{j \in M_1} \tilde{a}_{ij} + \sum_{j \in N_1} \tilde{b}_{ij} = \tilde{d}_i \tag{7}$$

where $f_i \in (0, 1)$, M_0 and M_1 represent the sets of 0-1 variables that are nonbasic at lower and upper bounds respectively in the current tableau, N_0 and N_1 represent the sets of continuous variables that are nonbasic at lower and upper bounds in the current tableau, and $B(i)$ represents the index of the ith basic variable. We would like to generate a strong cut from (7).

In (7) we assume that the coefficients $\tilde{a}_{ij}$ and $\tilde{b}_{ij}$ are positive. If not, we complement the corresponding variables (they will therefore switch from M_0 to M_1, N_0 to N_1 and vice-versa). Moreover, if we relax (7) to an inequality, we obtain the knapsack constraint *in standard form*

$$x_{B(i)} + \sum_{j \in M_0 \cup M_1} \tilde{a}_{ij} x_j + \sum_{j \in N_0 \cup N_1} \tilde{b}_{ij} y_j \leq \tilde{d}_i \tag{8}$$

which defines a mixed 0-1 knapsack polytope that we call PS_i. We can also relax the equality in the other direction and still obtain a mixed 0-1 knapsack polytope in standard form by complementing all the variables.

Note that we can fix to 0 all the 0-1 variables, including $x_{B(i)}$ whose coefficients in PS_i are bigger than $\tilde{d}_i$. By doing so, we detect that PQ is not full-dimensional and obtain some members of its equality set. Among the inequalities that can possibly be generated in this way, the only one that cuts off the current solution of the LP relaxation is $x_{B(i)} \leq 0$. So we will assume throughout this paper that variables for which the previous discussion applies have been substituted out of (8), that $x_{B(i)}$ is still present in (8) and that PS_i is full-dimensional. In $PS_i(M_0, M_1, N_0, N_1)$, the relaxed tableau row in standard form reads

$$x_{B(i)} \leq f_i \tag{9}$$

with $0 < f_i < 1$. Since $x_{B(i)}$ is an integer variable, the inequality

$$x_{B(i)} \leq 0 \tag{10}$$

is valid for $PS_i(M_0, M_1, N_0, N_1)$ and is clearly violated by the current solution. Note that (10) can be extended to a facet of PS_i by sequentially lifting all the nonbasic variables. Although all lifting orders provide such an inequality, some orders may lead to complex lifting schemes that will usually be expensive to carry out. So we will settle for simple schemes. Note that the mixed integer solutions of Q satisfy the constraints defining PS_i so that the inequalities we obtain are valid for PQ. Also, since all the nonbasic variables are fixed at their upper or lower bounds, the inequality obtained at the end of the lifting process will be violated and its absolute violation will be f_i. In order to implement this scheme, we need to lift both continuous and 0-1 variables. The lifting of 0-1 variables from a knapsack inequality is well-known, see for example Gu, Nemhauser and Savelsbergh [10]. For the lifting of continuous variables, we will use the results developed in [18,19]. We only need to consider the lifting of continuous variables fixed at 0 (lifting from 0) and fixed at 1 (lifting from 1) since all the continuous variables are nonbasic.

3   A Family of Tableau Cuts

In this section we show that we can choose the lifting order in a way that makes the whole lifting sequence polynomial. Consider PS_i, the mixed 0-1 knapsack polytope in standard form associated with a row of the tableau whose basic variable $x_{B(i)}$ is fractional. First define $M_0^< = \{l \in M_0 \,|\, \tilde{a}_{il} \le f_i + \sum_{j \in M_1} \tilde{a}_{ij}\}$ and $M_0^> = \{l \in M_0 \,|\, f_i + \sum_{j \in M_1} \tilde{a}_{ij} < \tilde{a}_{il} \le \tilde{d}_i\}$. Note that M_0 is completely covered by $M_0^<$ and $M_0^>$ since PS_i is full-dimensional. We will lift the variables in the order M_1, $M_0^<$, N_1, N_0 and $M_0^>$. We discuss the lifting problems next and illustrate them on the following example.

Example 1. Consider

$$\max \quad x_1 + 2x_2 + 3x_3 + 4x_4 + 10y_1$$
$$\text{s.t. } 10x_1 + 7x_2 + 6x_3 + 2x_4 + 2y_1 \le 17$$
$$6x_1 + 12x_2 + 13x_3 + 7x_4 + 13y_1 \le 32.$$

We introduce the nonnegative continuous slacks $\tilde{y}_2$ and $\tilde{y}_3$. They are bounded above by 17 and 32 respectively and below by 0. We use these bounds to rescale $\tilde{y}_2$ and $\tilde{y}_3$ so that their domains correspond to the interval $[0, 1]$. We call the scaled variables y_2 and y_3. We solve the linear relaxation of this problem using the simplex method. The optimal tableau is given by

$$\frac{96}{13} + \max - \frac{5}{13}x_1 - \frac{10}{13}x_2 \qquad + \frac{31}{13}x_4 + \frac{91}{13}y_1 \qquad - \frac{96}{13}y_3$$
$$\text{s.t.} \quad \frac{6}{13}x_1 + \frac{12}{13}x_2 + x_3 + \frac{7}{13}x_4 + y_1 \qquad + \frac{32}{13}y_3 = \frac{32}{13}$$
$$\frac{94}{221}x_1 + \frac{19}{221}x_2 \qquad - \frac{16}{221}x_4 - \frac{52}{221}y_1 + y_2 - \frac{192}{221}y_3 = \frac{29}{221}$$

where the variables x_3 and y_2 are basic, the variables x_1, x_2 and y_3 are nonbasic at lower bound, and the variables x_4 and y_1 are nonbasic at upper bound. The current objective value is $\frac{218}{13}$. For ease of exposition, we present the tableau in fractional form, although our analysis does not require it.

We derive a cut from the first row of the tableau whose basic variable $x_3 = \frac{12}{13}$. All the coefficients of this row are nonnegative, so this row is already in standard form. The polytope from which we generate our cuts is defined by the inequality

$$\frac{6}{13}x_1 + \frac{12}{13}x_2 + x_3 + \frac{7}{13}x_4 + y_1 + \frac{32}{13}y_3 \le \frac{12}{13} + \frac{7}{13} + 1.$$

Moreover, we have that $M_0^< = \{1, 2\}$, $M_0^> = \emptyset$, $M_1 = \{4\}$, $N_0 = \{3\}$ and $N_1 = \{1\}$. $\qquad\square$

3.1   Lifting the Members of M_1

First we lift the variables of M_1 in (10). The following theorem is a corollary of well-known results on lifting 0-1 variables.

Theorem 2. *Assume $M_1 = \{1, \ldots, s\}$ and let $\mu_0 = 1 - f_i$. For $j = 1, \ldots, s$, let $\alpha_j = 0$ and $\mu_j = \mu_{j-1} - \tilde{a}_{ij}$ when $\tilde{a}_{ij} < \mu_{j-1}$ and let $\alpha_j = 1$ and $\mu_j = \mu_{j-1}$ otherwise, then*

$$x_{B(i)} + \sum_{j=1}^{s} \alpha_j x_j \leq \sum_{j=1}^{s} \alpha_j \tag{11}$$

is a facet of $PS_i(M_0, \emptyset, N_0, N_1)$. □

Inequality (11) is obtained by sequentially lifting the variables of M_1 in (10) and is a cover inequality since $\alpha_j \in \{0,1\}$. Moreover Theorem 2 yields a polynomial algorithm $O(n)$ to obtain (11). We will assume from now on that at least one variable of M_1 is lifted at value 1. The case in which no variable from M_1 yields $\alpha = 1$ is discussed in Section 3.6.

Example 1 (continued). We lift x_4 from 1 by computing $\mu = 1/13$. Since $\tilde{a}_{14} = \frac{7}{13} \geq \frac{1}{13}$, we have $\alpha_4 = 1$. So $x_3 + x_4 \leq 1$ is a facet of $PS_i(M_0, \emptyset, N_0, N_1)$. □

3.2 Lifting Members of $M_0^<$

The minimal cover obtained in Section 3.1 is next lifted with respect to the variables of $M_0^<$. There is a polynomial algorithm to perform this task, see Nemhauser and Wolsey [15] and Zemel [21]. We will not present it here, but we note that all the lifting coefficients of the members of $M_0^<$ can be computed in time $O(n^2)$ and we denote the resulting facet-defining inequality of $PS_i(M_0^>, \emptyset, N_0, N_1)$ by

$$\sum_{j \in M \setminus M_0^>} \alpha_j x_j \leq \delta. \tag{12}$$

Example 1 (continued). We lift the variables x_1 and x_2 in this order and we determine that both of these lifting coefficients are 0. Therefore $x_3 + x_4 \leq 1$ is a facet of $PS_i(M_0^>, \emptyset, N_0, N_1) = PS_i(\emptyset, \emptyset, N_0, N_1)$. □

3.3 Lifting Members of N_1

For the lifting from 1 of continuous variables in a 0-1 inequality, we have given a pseudo-polynomial lifting scheme [18] that is based on the function Λ defined, from (12), as

$$\Lambda(w) = \min \sum_{j \in M \setminus M_0^>} \tilde{a}_{ij} x_j - \left(f_i + \sum_{j \in M_1} \tilde{a}_{ij} \right)$$

$$\text{s.t.} \sum_{j \in M \setminus M_0^>} \alpha_j x_j = \delta + w$$

$$x_j \in \{0,1\} \quad \forall j \in M \setminus M_0^>.$$

If, for some w, the problem defining $\Lambda(w)$ is infeasible, we define $\Lambda(w) = \infty$. We say that the function Λ is superlinear if $w^*\Lambda(w) \geq w\Lambda(w^*)$ for all $w \geq w^*$ where $w^* = \max \operatorname{argmin}\{\Lambda(w) \mid w > 0\}$. When the function Λ is superlinear, the lifting algorithm proposed in [18] can be significantly simplified. The following theorem establishes that the lifted covers obtained in Section 3.2 are superlinear with $w^* = 1$.

Theorem 3 ([19]). *For (12) we have $\Lambda(w) \geq w\Lambda(1)$, for all $w > 0$.*   □

This property leads to the following lifting theorem.

Theorem 4 ([19]). *Assume $N_1 = \{1, \ldots, s\}$, then*

$$\sum_{j \in M \setminus M_o^{>}} \alpha_j x_j + \theta \sum_{j=p+1}^{s} \tilde{b}_{ij} y_j \leq \delta + \theta \sum_{j=p+1}^{s} \tilde{b}_{ij} \tag{13}$$

is a facet of $PS_i(M_0^{>}, \emptyset, N_0, \emptyset)$ where $p = \max\{k \in \{1, \ldots, s\} \mid \sum_{j=1}^{k} \tilde{b}_{ij} < \Lambda(1)\}$, $\theta = \frac{1}{\Lambda(1)-b}$ and $b = \sum_{j=1}^{p} \tilde{b}_{ij}$.   □

This refined version of the lifting algorithm, proven in [19], is valid for lifted covers and requires only the knowledge of $\Lambda(1)$. This value can be computed in time $O(n^2)$. So, once $\Lambda(1)$ is known, Theorem 4 yields a way to compute all the lifting coefficients of variables in N_1 in time $O(n)$ independent of the lifting order. Observe that it is possible that $p = s$ in which case no continuous variable appears in (13).

Example 1 (continued). We lift from 1 the continuous variable y_1. Note that $1 > \Lambda(1) = \frac{1}{13}$. Therefore the inequality $x_3 + x_4 + 13y_1 \leq 14$ is a facet of $PS_i(\emptyset, \emptyset, N_0, \emptyset)$.   □

3.4 Lifting Members of N_0

At this stage, we lift the members of N_0. We have shown in [18] that the lifting coefficient is 0 almost always. More generally,

Theorem 5 ([18]). *Assume that (13) is a valid inequality of $PS(M_0^{>}, \emptyset, N_0, \emptyset)$ and that $k \in N_0$. Assume there exists $x^* \in S(M_0^{>}, \emptyset, N_0, \emptyset)$ that satisfies (13) at equality and the defining inequality of $PS_i(M_0^{>}, \emptyset, N_0, \emptyset)$ strictly at inequality, then in lifting y_k from 0, the lifting coefficient is 0.*   □

For the inequality obtained in Theorem 4, the point $x^*_{B(i)} = 0$, $x^*_j = 0$ for all $j \in M_0^{<}$ and $x^*_j = 1$ for all $j \in M_1$ satisfies the assumption of Theorem 5. Therefore the lifting coefficients for all members of N_0 are 0. It follows that inequality (13) is a facet of $PS(M_0^{>}, \emptyset, \emptyset, \emptyset)$.

Example 1 (continued). We lift the continuous variable y_3. Theorem 5 implies that $x_3 + x_4 + 13y_1 \leq 14$ is a facet of $PS_i = PS_i(\emptyset, \emptyset, \emptyset, \emptyset)$. It is also a cut for our initial problem. Moreover, it can be verified that it is a facet of PQ because the points $(x_1, x_2, x_3, x_4, y_1) = (0,0,0,1,1), (0,0,1,0,1), (0,1,0,1,1),$ $(1,0,0,1,1)$ and $(0,0,1,1,\frac{12}{13})$ belong to Q, make the inequality tight and are affinely independent. If we add this cut and reoptimize, the solution we obtain is $(0,0,1,1,\frac{12}{13})$, which is optimal and has an objective value of $\frac{211}{13}$.   □

3.5   Lifting Members of $M_0^>$

Finally we lift the members of $M_0^>$, i.e. the variables with large coefficients. First suppose the inequality we lift is a 0-1 lifted cover, i.e. $p = s$. Again, the dynamic programming algorithm presented in [15,21] can be used and all the lifting coefficients for the members of $M_0^>$ can be computed in time $O(n^2)$. Now suppose that $p < s$. There is a closed form expression for the lifting of members of $M_0^>$. This closed form expression developed in [19] for general superlinear inequalities is described in the next theorem. For $a \in \mathbb{R}$, we define $(a)^+ = \max\{a, 0\}$.

Theorem 6 ([19]). *Assume $p < s$ and (13) is a facet of $PS_i(M_0^>, \emptyset, \emptyset, \emptyset)$. Then*

$$\sum_{j \in M \backslash M_0^>} \alpha_j x_j + \theta \sum_{j=p+1}^{s} \tilde{b}_{ij} y_j + \sum_{j \in M_0^>} G(a_j) x_j \leq \delta + \theta \sum_{j=p+1}^{s} \tilde{b}_{ij} \qquad (14)$$

where $G(a) = \delta + \theta(a - d_i^* - b)^+$ and $d_i^* = \tilde{d}_i - \sum_{j \in N_1} \tilde{b}_{ij}$ is a facet of PS_i.   □

Theorem 6 leads to a linear time algorithm for the lifting of members of $M_0^>$ when $p < s$. Thus, the cut we add to the current LP relaxation of the 0-1 mixed integer program is either a 0-1 lifted cover or (14). The previous discussion shows that, in either case, this cut can be derived in time $O(n^2)$.

3.6   Final Remarks on Lifting

In the previous discussion, we omitted the case where all the lifting coefficients of the variables of M_1 are 0. In this case, all the lifting coefficients of members of $M_0^<$ are 0 too because the inequalities $x_j \leq 0$ for $j \in M_0^<$ would be valid in $PS_i(M_0^>, \emptyset, N_0, N_1)$ which contradicts the definition of $M_0^<$. At least one member of N_1 is lifted with a positive coefficient, otherwise $x_{B(i)} \leq 0$ would be valid for PS_i which contradicts the full-dimensionality of PS_i. The inequality we obtain is a facet of $PS_i(M_0^>, \emptyset, N_0, \emptyset)$ that can be turned into a facet of PS_i using Theorems 5 and 6. So our cut generation procedure returns either a facet of PS_i, if it is full dimensional, or, as discussed in Section 2, a member of its equality set, if it is not full dimensional.

All the cuts are generated in time $O(n^2)$ and are of the standard form

$$x_{B(i)} + \sum_{j \in M_0 \cup M_1} \alpha_j x_j + \sum_{j \in N_0 \cup N_1} \beta_j y_j \leq \sum_{j \in M_1} \alpha_j + \sum_{j \in N_1} \beta_j \qquad (15)$$

with $\alpha_j \geq 0$ for $j \in M_0 \cup M_1$ and $\beta_j \geq 0$ for $j \in N_0 \cup N_1$. This simple observation leads to the following proposition that will be used to establish finite convergence of our algorithm. For convenience, we will now incorporate the upper bounds on variables in the set of constraints before using the simplex method. We refer to this variant of the simplex method as being *without upper bounds*.

Proposition 1. *If we apply the simplex method without upper bounds, the cut generated from the simplex tableau row* (7) *where* $0 < f_i < 1$ *is of the form*

$$x_{B(i)} + \sum_{j \in M_0} \alpha_j x_j + \sum_{j \in N_0} \beta_j y_j \leq 0 \qquad (16)$$

where $\alpha_j \leq 0$ *if* $\tilde{a}_{ij} < 0$, $\alpha_j \geq 0$ *if* $\tilde{a}_{ij} > 0$, $\beta_j \leq 0$ *if* $\tilde{b}_{ij} < 0$, *and* $\beta_j \geq 0$ *if* $\tilde{b}_{ij} > 0$.

Proof. The tableau row (7) on which we generate the cut contains only M_0 and N_0. After all the variables with negative coefficients in (7) are complemented to fit in the standard format, we generate (15) that has only nonnegative coefficients. Since all the members of M_1 and N_1 are complemented variables, they need to be complemented back yielding (16).   □

4 A Finitely Convergent Algorithm for 0-1 Mixed Integer Programming

The ability to generate a cut from every row of the simplex tableau where a basic integer variable is fractional is reminiscent of Gomory cuts. Now, as done by Gomory, we prove a finite convergence result. The approach we take is similar to the one described by Nourie and Venta [17]. Consider the mixed integer problem (1), (3), (5) and (6). For ease of notation, in this section, we denote all the variables in the 0-1 mixed integer problem by x, even if they are continuous, i.e. we define $x_{m+i} = y_i$ for $i = 1, \dots, n + h$. We assume that every extreme point of PQ say x^q is such that $cx^q \in \mathbb{Z}$. This condition can always be met by adequately scaling the objective function. We say that $u \in \mathbb{R}^t$ is lexicographically larger than 0 ($u \succ 0$) if there exists $k \in \{1, \dots, t\}$ such that $u_1 = \dots = u_{k-1} = 0$ and $u_k > 0$. We say also that, for $u, v \in \mathbb{R}^t$, u is lexicographically larger than v ($u \succ v$) if $u - v \succ 0$. If the two vectors we compare are of different length, we just drop the last components of the longer one and perform the comparison on vectors of the same size.

We modify the fractional cutting plane algorithm (*FCPA*) presented in [15], p. 368-369 to handle our cuts instead of Gomory's and prove its convergence using the arguments presented in [15], p. 370-373. We recall some of the assumptions under which convergence is established.

(i)   We use the simplex method without upper bounds.
(ii) The objective function is restricted to be integer. This is a valid since we know that cx^q is integer for every extreme point x^q of PQ. Therefore, an equality of the form $x_0 = cx$, where x_0 is an integer variable, is introduced in the set of constraints and the objective function is replaced by x_0.

(iii) We solve the linear relaxations in such a way that the solution we obtain is lexicographically maximum for the set of optimal solutions and we include rows of the form $x_j - x_j = 0$ in the tableaux for nonbasic variables.

(iv) We generate a single cut at every step of the algorithm and we generate it from the simplex tableau row whose basic integer variable is fractional and has the smallest index.

(v) After we add a cut, the problem must still be of the initial form. In our case, it suffices to introduce the slack and rescale it to be in the interval $[0, 1]$.

The fact that the variable x_0 is a general integer variable is not a problem in the lifting steps of our algorithm since we can always keep x_0 basic. Note that when x_0 is fractional, say $x_0 = q$, the cut $x_0 \leq \lfloor q \rfloor$ is valid. From the previous observation and Proposition 1 we conclude that the cut we generate from the kth simplex tableau row with basic variable x_k

$$x_k + \sum_{j \in NB} \tilde{a}_{kj} x_j = \tilde{d}_k \tag{17}$$

is of the form

$$\alpha_k x_k + \sum_{j \in NB} \alpha_j x_j \leq \delta_k \tag{18}$$

where $\alpha_k = 1$, $\delta_k = \lfloor \tilde{d}_k \rfloor$ and NB is the set of nonbasic variables in the current simplex tableau.

Having just presented a scheme in which we can embed our cuts, we now prove that these cuts are strong enough to yield an optimal solution in a finite number of iterations. It is not true that this property will be achieved by any family of violated inequalities. For example, as shown by Gomory and Hoffman [9], the family of cuts that require the sum of the nonbasic integer variables to be at least one, see Dantzig [6], do not necessarily lead to a convergent algorithm. They need to be improved, as described by Bowman and Nemhauser [4], to yield a convergent algorithm. A sufficient condition on the strength of cuts needed to obtain finite convergence is described in the next proposition.

Proposition 2. *Assume that all the cuts generated from simplex tableau rows (17) are of the form (18). Let f_k be the fractional part of $\tilde{d}_k$ and assume that for every $l \in NB$ such that $\alpha_l - \alpha_k \tilde{a}_{kl} < 0$ and $\tilde{a}_{kl} \geq 0$, we have*

$$f_k \alpha_l + \tilde{a}_{kl}(\alpha_k \lfloor \tilde{d}_k \rfloor - \delta_k) \geq 0. \tag{19}$$

Then FCPA converges in a finite number of iterations.

Proof. We extend the proof of Proposition 3.7 from [15]. Assume that we have already added t cuts. We work with a simplex tableau that has $v(t) = m+n+h+t$ rows of the form of (17). Assume that x^t is an optimal solution of the current relaxation and that k is the smallest index among all 0-1 variables that are currently fractional. We define $S^t = (x_0^t, \ldots, x_{k-1}^t, \lfloor x_k^t \rfloor, u_{k+1}, \ldots, u_m)$, where $u_{k+1}, \ldots, u_m$ are upper bounds on the integer variables, i.e. 1 in our case. We have that $x^t \succ S^t$. We need to prove $x^{t+1} \preceq S^t$. Assume that, from row k of the tableau, we generate the cut $\alpha_k x_k + \sum_{j \in NB} \alpha_j x_j + u x_{v(t+1)} = \delta_k$, where NB is the set of nonbasic variables, and add it to the current formulation. Note that we introduce u (which can always be chosen to be positive) as a way to rescale the slacks since their domain has to be the interval $[0,1]$. After adding the cut, we make the slack basic and therefore obtain the basic, primal infeasible, dual optimal tableau

$$x_p + \sum_{j \in NB} \tilde{a}_{pj} x_j = \tilde{d}_p \qquad \forall p \in \{1, \ldots, v(t)\}$$

$$x_{v(t+1)} + \sum_{j \in NB} \frac{\alpha_j - \alpha_k \tilde{a}_{kj}}{u} x_j = \frac{\delta_k - \alpha_k \tilde{d}_k}{u}$$

in which the column associated with x_j, $a_j \succ 0 \forall j \in NB$. Let $(\hat{x}_0^t, \hat{x}^t)$ be the basic solution obtained after a single dual simplex pivot and let x_l be the variable that becomes basic. Clearly $\beta_l = \frac{\alpha_l - \alpha_k \tilde{a}_{kl}}{u} < 0$, $\frac{\delta_k - \alpha_k \tilde{d}_k}{u} < 0$ and we have

$$\begin{pmatrix} \hat{x}_0^t \\ \hat{x}^t \end{pmatrix} = \begin{pmatrix} x_0^{t-1} \\ x^{t-1} \end{pmatrix} - \frac{\delta_k - \alpha_k \tilde{d}_k}{\alpha_l - \alpha_k \tilde{a}_{kl}} a_l.$$

We have that $a_l \succ 0$ and $\frac{\delta_k - \alpha_k \tilde{d}_k}{\alpha_l - \alpha_k \tilde{a}_{kl}} > 0$. Now let r be the minimum index for which $\tilde{a}_{rl} > 0$. We distinguish two cases. First $r \leq k - 1$. In that case $\hat{x}_j^t = x_j^t$ for all $j < r$ and $\hat{x}_r^t < x_r^t$. Therefore $\hat{x}^t \prec S^t$ and so $x^{t+1} \prec S^t$. Now assume $r \geq k$. We have $\hat{x}_j^t = x_j^t$ for all $j < k$ and

$$\hat{x}_k^t = \lfloor \tilde{d}_k \rfloor + \frac{f_k \alpha_l + \tilde{a}_{kl}(\alpha_k \lfloor \tilde{d}_k \rfloor) - \delta_k)}{\alpha_l - \alpha_k \tilde{a}_{kl}}.$$

Now since $\tilde{a}_l \succ 0$ and $\tilde{a}_{jl} = 0$ for $j < k$, we have that $\tilde{a}_{kl} \geq 0$. Using (19), we conclude that the numerator of the fraction is nonnegative and, since its denominator is negative, that $\hat{x}_k^t \leq \lfloor \tilde{d}_k \rfloor$. It follows that $x^{t+1} \preceq \hat{x}_k^t \preceq S^t$. $\qquad \square$

We use Proposition 2 to prove that our algorithm is finite.

Theorem 7. *There exists a pure cutting plane algorithm, based on the cuts presented in Section 3, that solves Problem (1), (3), (5) and (6) in a finite number of iterations.*

Proof. According to (18), we have $\alpha_k = 1$ and $\delta_k = \lfloor \tilde{d}_k \rfloor$. Condition (19) becomes $f_k \alpha_l \geq 0$ for all $l \in NB$ such that $\alpha_l - \tilde{a}_{kl} < 0$ and $\tilde{a}_{kl} \geq 0$. Since $\tilde{a}_{kl} \geq 0$, we know from Proposition 1 that $\alpha_l \geq 0$ and so we conclude that condition (19) is satisfied since $f_k > 0$. $\qquad \square$

Proposition 2 can also be used to show that Gomory cuts for integer programs yield a finitely convergent algorithm. Starting from the simplex tableau row (17), Gomory cuts are of the form $\sum_{j \in NB} f_j z_j \geq g_k$ where f_j and g_k are the fractional parts of $\tilde{a}_{kj}$ and $\tilde{d}_k$. Therefore we have that $\alpha_k = 0$, $\delta_k = -g_k$ and $\alpha_l = -f_l$. Condition (19) is then $-g_k f_l + \tilde{a}_{kl} g_k = g_k(\tilde{a}_{kl} - f_l) \geq 0$ which is satisfied when $\tilde{a}_{kl} \geq 0$ and $f_l > 0$ because g_k is positive.

5   Conclusions

The cuts we have developed are not strictly comparable to Gomory cuts. We first relax a simplex tableau row into a 0-1 mixed integer knapsack and then we find a facet of this knapsack relaxation. Thus our cuts are strong since they are facets of a good relaxation. But it is interesting that they are also robust in that they can be implemented to yield a finite pure cutting plane algorithm. Nevertheless, the practical use of these cuts is likely to come from imbedding them in a branch-and-cut algorithm which we are currently developing.

References

1. E. Balas. Facets of the knapsack polytope. *Mathematical Programming*, 8:146–164, 1975.
2. E. Balas, S. Ceria, and G. Cornuéjols. A lift-and-project cutting plane algorithm for mixed 0-1 programs. *Mathematical Programming*, 58:295–324, 1993.
3. E. Balas and E. Zemel. Facets of the knapsack polytope from minimal covers. *SIAM Journal on Applied Mathematics*, 34:119–148, 1978.
4. V.J. Bowman and G.L. Nemhauser. A finiteness proof for modified Dantzig cuts in integer programming. *Naval Research Logistics Quarterly*, 17:309–313, 1970.
5. H.P. Crowder, E.L. Johnson, and M.W. Padberg. Solving large-scale zero-one linear programming problems. *Operations Research*, 31:803–834, 1983.
6. G.B. Dantzig. Note on solving linear programs in integers. *Naval Research Logistics Quarterly*, 6:75–76, 1959.
7. R.E. Gomory. Outline of an algorithm for integer solutions to linear programs. *Bulletin of the American Mathematical Society*, 64:275–278, 1958.
8. R.E. Gomory. An algorithm for the mixed integer problem. Technical Report RM-2597, RAND Corporation, 1960.
9. R.E. Gomory and A.J. Hoffman. On the convergence of an integer programming process. *Naval Research Logistics Quarterly*, 10:121–124, 1963.
10. Z. Gu, G.L. Nemhauser, and M.W.P. Savelsbergh. Lifted cover inequalities for 0-1 integer programs: Complexity. *INFORMS Journal on Computing*, 11:117–123, 1999.
11. P.L. Hammer, E.L. Johnson, and U.N. Peled. Facets of regular 0-1 polytopes. *Mathematical Programming*, 8:179–206, 1975.
12. E.L. Johnson, G.L. Nemhauser, and M.W.P. Savelsbergh. Progress in linear programming based branch-and-bound algorithms:an exposition. *INFORMS Journal on Computing*, 12:2–23, 2000.
13. H. Marchand, A. Martin, R. Weismantel, and L. Wolsey. Cutting planes in integer and mixed integer programming. Technical Report 9953, Université Catholique de Louvain, 1999.

14. H. Marchand and L.A. Wolsey. The 0-1 knapsack problem with a single continuous variable. *Mathematical Programming*, 85:15–33, 1999.
15. G.L. Nemhauser and L.A. Wolsey. *Integer and Combinatorial Optimization*. Wiley, New York, 1988.
16. G.L. Nemhauser and L.A. Wolsey. A recursive procedure for generating all cuts for 0-1 mixed integer programs. *Mathematical Programming*, 46:379–390, 1990.
17. F.J. Nourie and E.R. Venta. An upper bound on the number of cuts needed in Gomory's method of integer forms. *Operations Research Letters*, 1:129–133, 1982.
18. J.-P. P. Richard, I.R. de Farias, and G.L. Nemhauser. Lifted inequalities for 0-1 mixed integer programming : Basic theory and algorithms. Technical Report 02-05, Georgia Institute of Technology, 2002.
19. J.-P. P. Richard, I.R. de Farias, and G.L. Nemhauser. Lifted inequalities for 0-1 mixed integer programming : Superlinear lifting. Technical report, Georgia Institute of Technology, 2002. (in preparation).
20. L.A. Wolsey. Faces for a linear inequality in 0-1 variables. *Mathematical Programming*, 8:165–178, 1975.
21. E. Zemel. Easily computable facets of the knapsack polytope. *Mathematics of Operations Research*, 14:760–764, 1989.

Mixed-Integer Value Functions in Stochastic Programming

Rüdiger Schultz

Institute of Mathematics, Gerhard-Mercator University Duisburg
Lotharstr. 65, D-47048 Duisburg, Germany
schultz@math.uni-duisburg.de

Abstract. We discuss the role of mixed-integer value functions in the theoretical analysis of stochastic integer programs. It is shown how the interaction of value function properties with basic results from probability theory leads to structural statements in stochastic integer programming.

1 Stochastic Integer Programs

Stochastic programming models are deterministic equivalents to random optimization problems. In the present paper we confine ourselves to linear two-stage models involving integer requirements. The random optimization problem behind these models reads as follows

$$\min_{x,y,y'} \{c^T x + q^T y + q'^T y' \; : \; Tx + Wy + W'y' = h(\omega),$$

$$x \in X, \, y \in \mathbb{Z}_+^{\bar{m}}, \, y' \in \mathbb{R}_+^{m'}\}. \tag{1}$$

We assume that all ingredients above have conformal dimensions, that W, W' are rational matrices, and that $X \subseteq \mathbb{R}^m$ is a nonempty closed polyhedron, possibly involving integrality constraints on components of the vector x.

Together with (1) we have a scheme of alternating decision and observation: The decision on x has to be made prior to observing the outcome of the random vector $h(\omega)$, and the vector (y, y') is selected only after having decided on x and observed $h(\omega)$. This setting corresponds to a variety of practical optimization problems under uncertainty. It readily extends to the multi-stage situation where finitely (or even infinitely) many of the above alternations occur, see [6,12,17] for further details on stochastic programming modelling.

As a mathematical object, problem (1) is ill-posed, since at the moment of decision on x it is not clear which vectors x are feasible, let alone optimal. As a remedy, let us proceed as follows. Rewrite (1) by separating the optimizations in x and (y, y'):

$$\min_x \left\{ c^T x + \min_{y,y'} \{q^T y + q'^T y' : Wy + W'y' = h(\omega) - Tx, \right.$$

$$\left. y \in \mathbb{Z}_+^{\bar{m}}, y' \in \mathbb{R}_+^{m'}\} \; : \; x \in X \right\}. \tag{2}$$

M. Jünger et al. (Eds.): Combinatorial Optimization (Edmonds Festschrift), LNCS 2570, pp. 171–184, 2003.

This is where the mixed-integer value function enters the scene. Indeed, in (2) we have an inner optimization problem with right-hand side parameter $h(\omega) - Tx$. Introducing the mixed-integer value function

$$\Phi(t) := \min\{q^T y + q'^T y' \; : \; Wy + W'y' = t, \; y \in \mathbb{Z}_+^{\bar{m}}, \; y' \in \mathbb{R}_+^{m'}\}, \quad (3)$$

(2) turns into

$$\min_x\{c^T x + \Phi(h(\omega) - Tx) \; : \; x \in X\}. \quad (4)$$

In this way, we obtain the family $f(x, \omega) := c^T x + \Phi(h(\omega) - Tx)$, $x \in X$ of real-valued random variables. The problem is still ill-posed, since in (4) the meaning of "$\min_x$" remains unclear, i.e., it is still open how to select a "best" random variable $f(x, .)$. In stochastic programming, scalar parameters of $f(x, \omega)$, $x \in X$ provide criteria for making the "best" selection.

The most widely used scalar parameter in this respect is the expectation. Assuming that $h(\omega) \in \mathbb{R}^s$ is a random vector on some probability space $(\Omega, \mathcal{A}, \mathbb{P})$ we obtain the well-posed optimization problem

$$\min\left\{ \int_\Omega (c^T x + \Phi(h(\omega) - Tx))\, \mathbb{P}(d\omega) \; : \; x \in X \right\}. \quad (5)$$

In terms of the random optimization problem (1), the model (5) suggests to select, before observing $h(\omega)$, i.e., in a "here-and-now" manner, a decision x such that the expected value of the random costs $c^T x + \Phi(h(\omega) - Tx)$ becomes minimal.

When addressing risk aversion, other scalar parameters are useful. In the context of stochastic programming some first proposals have been made in [15,16,23]. Following the probability-based approach in [23] we introduce some threshold level $\varphi_o \in \mathbb{R}$ and consider minimization of the probability that the random costs $c^T x + \Phi(h(\omega) - Tx)$ exceed φ_o. This leads to the optimization model

$$\min\left\{ \mathbb{P}(\{\omega \in \Omega \; : \; c^T x + \Phi(h(\omega) - Tx) > \varphi_o\}) \; : \; x \in X \right\}. \quad (6)$$

As mathematical objects, (5) and (6) are optimization problems in x whose objectives we denote by $Q_{\mathbb{E}}(x)$ and $Q_{\mathbb{P}}(x)$, respectively. It is evident, that the mixed-integer value function Φ essentially determines the structure of the functions $Q_{\mathbb{E}}$ and $Q_{\mathbb{P}}$. As we will see, there is a fruitful interaction of properties of Φ with basic statements from probability theory.

The article is organized as follows. In Section 2 we report on what is known about the mixed-integer value function Φ. Section 3 aims at putting together value function properties with basic probability theory. Proceeding step by step, we draw conclusions from various convergence results of probability theory. The final section is an outlook towards related issues beyond the scope of the present paper.

2   Mixed-Integer Value Functions

Studying the value function Φ is part of what is usually referred to as stability analysis of optimization problems or parametric optimization. In this area of research the accent is on properties of optimal values and optimal solution sets seen as (multi-)functions of parameters arising in the underlying optimization problems. A variety of results starting from linear programs and leading into nonlinear programming and optimal control is available, see the recent monograph [8] and references therein.

In the above stochastic programming context, the value function determines integrands of suitable integrals. Therefore, it is crucial to have *global* knowledge about the functional dependence of the optimal value on the respective parameter. The typical situation in *nonlinear* parametric optimization, however, is that properties of optimal values and optimal solutions are available only *locally* around given parameters. The most comprehensible class for which global results exist are mixed-integer linear programs. This explains that, so far, the models discussed in Section 1 do not go beyond the mixed-integer linear case. The stability of mixed-integer linear programs has been studied in a series of papers by Blair and Jeroslow out of which we refer to [7], and in the monographs [2,3].

Before discussing the mixed-integer value function Φ from (3), let us have a quick look at its linear-programming counterpart:

$$\Phi_{lin}(t) := \min\{q'^T y' \; : \; W'y' = t, \; y' \in \mathbb{R}_+^{m'}\}. \tag{7}$$

If we assume that $W'(\mathbb{R}_+^{m'})$ is full-dimensional and that

$$\{u \in \mathbb{R}^s \; : \; {W'}^T u \le q'\} \neq \emptyset,$$

then the latter set has vertices d_k, $k = 1, \ldots, K$, and it holds by linear programming duality that

$$\Phi_{lin}(t) = \max\{t^T u : {W'}^T u \le q'\} = \max_{k=1,\ldots,K} d_k^T t \quad \text{for all } t \in W'(\mathbb{R}_+^{m'}).$$

Hence, Φ_{lin} is convex and piecewise linear on its (conical) domain of definiton. Without going into details, we mention that this convexity has far reaching consequences when setting up the expectation based stochastic programming model (5) in case integer requirements are missing in the random optimization problem (1), see [6,12,17] for structural and algorithmic results.

Let us now turn our attention to the mixed-integer case. In (3) we impose the basic assumptions that

$$W(\mathbb{Z}_+^{\bar{m}}) + W'(\mathbb{R}_+^{m'}) = \mathbb{R}^s \text{ and } \{u \in \mathbb{R}^s \; : \; W^T u \le q, \; {W'}^T u \le q'\} \neq \emptyset.$$

Then $\Phi(t)$ is a well-defined real number for all $t \in I\!R^s$, [14]. Moreover it holds

$$
\begin{aligned}
\Phi(t) &= \min\{q^T y + q'^T y' \ : \ Wy + W'y' = t, \ y \in \mathbb{Z}_+^{\bar{m}}, \ y' \in I\!R_+^{m'}\} \\
&= \min_y\{q^T y + \min_{y'}\{q'^T y' \ : \ W'y' = t - Wy, \ y' \in I\!R_+^{m'}\} \ : \ y \in \mathbb{Z}_+^{\bar{m}}\} \\
&= \min_y\{\Phi_y(t) \ : \ y \in \mathbb{Z}_+^{\bar{m}}\}, \tag{8}
\end{aligned}
$$

where

$$
\Phi_y(t) = q^T y + \max_{k=1,\dots,K} d_k^T(t - Wy) \qquad \text{for all } t \in Wy + W'(I\!R_+^{m'}).
$$

Here, d_k, $k = 1, \dots, K$ are the vertices of $\{u \in I\!R^s \ : \ W'^T u \le q'\}$, and we have applied the argument about Φ_{lin} from the purely linear case. For $t \notin Wy + W'(I\!R_+^{m'})$ the problem $\min_{y'}\{q'^T y' \ : \ W'y' = t - Wy, \ y' \in I\!R_+^{m'}\}$ is infeasible, and we put $\Phi_y(t) = +\infty$. It is convenient to introduce the notation $Y(t) := \{y \in \mathbb{Z}_+^{\bar{m}} \ : \ \Phi_y(t) < +\infty\}$.

According to (8) the value function Φ is made up by the pointwise minimum of a family of convex, piecewise linear functions whose domains of definition are polyhedral cones arising as shifts of the cone $W'(I\!R_+^{m'})$. By our basic assumption $W(\mathbb{Z}_+^{\bar{m}}) + W'(I\!R_+^{m'}) = I\!R^s$, the cone $W'(I\!R_+^{m'})$ is full-dimensional.

Some first conclusions about the continuity of Φ may be drawn from the above observations:

(i)   Suppose that $t \in I\!R^s$ does not belong to any boundary of any of the sets $Wy+W'(I\!R_+^{m'})$, $y \in \mathbb{Z}^{\bar{m}}$. Then the same is true for all points τ in some open ball B around t. Hence, $Y(\tau) = Y(t)$ for all $\tau \in B$. With an enumeration $(y_n)_{n\in I\!N}$ of $Y(t)$ we consider the functions $\Phi^\kappa(\tau) := \min\{\Phi_{y_n}(\tau) : n \le \kappa\}$ for all $\tau \in B$. Then $\lim_{\kappa\to\infty} \Phi^\kappa(\tau) = \Phi(\tau)$ for all $\tau \in B$. Since, for any function Φ_y, its "slopes" are determined by the same, finitely many vectors d_k, $k = 1, \dots, K$, the functions $\Phi^\kappa, \kappa \in I\!N$ are all Lipschitz continuous on B with a uniform Lipschitz constant. Thus, the family of functions $\Phi^\kappa, \kappa \in I\!N$ is equicontinuous on B and has a pointwise limit there. Consequently, this pointwise limit Φ is continuous on B, in fact Lipschitz continuous with the mentioned uniform constant.

(ii) Any discontinuity point of Φ must be located at the boundary of some set $Wy + W'(I\!R_+^{m'})$, $y \in \mathbb{Z}^{\bar{m}}$. Hence, the set of discontinuity points of Φ is contained in a countable union of hyperplanes. Since $W'(I\!R_+^{m'})$ has only finitely many facets, this union of hyperplanes subdivides into finitely many classes, such that, in each class, the hyperplanes are parallel. By the rationality of the matrices W and W', within each class, the pairwise distance of the hyperplanes is uniformly bounded below by some positive number.

(iii) Let $t_n \to t$ and $y \in \mathbb{Z}^{\bar{m}}$ such that $t_n \in Wy + W'(I\!R_+^{m'})$ for all sufficiently large n. Since the set $Wy + W'(I\!R_+^{m'})$ is closed, this yields $t \in$

$Wy + W'(I\!R_+^{m'})$. Therefore, for sufficiently large n, $Y(t_n) \subseteq Y(t)$. This paves the way for showing that $\liminf_{t_n \to t} \Phi(t_n) \geq \Phi(t)$, which is the lower semicontinuity of Φ at t.

The above analysis has been refined in [2,3,7]. In particular, it is shown in Theorem 3.3 of [7] that, for each $t \in I\!R^s$, the minimization in (8) can be restricted to a finite set (depending on t, in general). Lemma 5.6.1. and Lemma 5.6.2. of [2] provide the representation of the continuity sets of Φ to be displayed in the subsequent proposition. The global proximity result in part (iv) of the subsequent proposition is derived in Theorem 1 at page 115 of [3] and in Theorem 2.1 of [7]. Altogether, we have the following statement about the mixed-integer value function Φ.

Proposition 1. *Let W, W' be matrices with rational entries and assume that $W(\mathbb{Z}_+^{\bar{m}}) + W'(I\!R_+^{m'}) = I\!R^s$ as well as $\{u \in I\!R^s : W^T u \leq q, W'^T u \leq q'\} \neq \emptyset$. Then it holds*

(i)   *Φ is real-valued and lower semicontinuous on $I\!R^s$,*

(ii) *there exists a countable partition $I\!R^s = \cup_{i=1}^{\infty} \mathcal{T}_i$ such that the restrictions of Φ to $\mathcal{T}_i$ are piecewise linear and Lipschitz continuous with a uniform constant $L > 0$ not depending on i,*

(iii) *each of the sets $\mathcal{T}_i$ has a representation $\mathcal{T}_i = \{t_i + \mathcal{K}\} \setminus \cup_{j=1}^{N} \{t_{ij} + \mathcal{K}\}$ where $\mathcal{K}$ denotes the polyhedral cone $W'(I\!R_+^{m'})$ and t_i, t_{ij} are suitable points from $I\!R^s$, moreover, N does not depend on i,*

(iv) *there exist positive constants β, γ such that $|\Phi(t_1) - \Phi(t_2)| \leq \beta \|t_1 - t_2\| + \gamma$ whenever $t_1, t_2 \in I\!R^s$.*

3   Implications of Probability Theory

Essential properties of the objective functions $Q_{I\!E}$ and $Q_{I\!P}$ in the optimization problems (5) and (6), respectively, have their roots in properties of Φ. We will study these interrelations by employing some basic tools from probability theory as can be found, for instance, in the textbooks [5,10,18].

Throughout this section we will impose the basic assumptions that W, W' are rational matrices, that $W(\mathbb{Z}_+^{\bar{m}}) + W'(I\!R_+^{m'}) = I\!R^s$, and that the set $\{u \in I\!R^s : W^T u \leq q, W'^T u \leq q'\}$ is non-empty.

For convenience we denote by μ the image measure $I\!P \circ h^{-1}$ on $I\!R^s$. With this notation, the functions $Q_{I\!E}$ and $Q_{I\!P}$ read

$$Q_{I\!E}(x) = \int_{I\!R^s} (c^T x + \Phi(h - Tx)) \, \mu(dh), \ x \in I\!R^m,$$

and

$$Q_{I\!P}(x) = \mu(\{h \in I\!R^s : c^T x + \Phi(h - Tx) > \varphi_o\}), \ x \in I\!R^m.$$

3.1   Measurability

Since $Q_{I\!E}(x)$ is esssentially determined by an integral over Φ and $Q_{I\!P}(x)$ involves a level set of Φ, it has to be assured that the integral and the probability are taken over a measurable function and a measurable set, respectively.

Proposition 2. *For any $x \in I\!R^m$, $f(x, h) := c^T x + \Phi(h - Tx)$ is a measurable function of h, implying in particular that $Q_{I\!P}(x)$ is well-defined for all $x \in I\!R^m$.*

   Proof: Φ being lower semicontinuous on $I\!R^s$, f is measurable as a superposition of measurable functions. Then, $\{h \in I\!R^s \; : \; f(x, h) > \varphi_o\}$ is a measurable subset of $I\!R^s$, and $Q_{I\!P}(x)$ is well-defined for all $x \in I\!R^m$.   q.e.d.

3.2   Integrability

A measurable function into the reals is called integrable, in case its positive and negative parts both are. Integrability is often established via an integrable majorant of the absolute value of the function in question. In the present context, integrability is important for assuring that $Q_{I\!E}(x)$ is well-defined for all $x \in I\!R^m$.

Proposition 3. *If μ has a finite first moment, i.e., if $\int_{I\!R^s} \|h\| \, \mu(dh) < \infty$, then $Q_{I\!E}(x)$ is well-defined for all $x \in I\!R^m$.*

   Proof: Our basic assumptions imply that $\Phi(0) = 0$. Together with Proposition 1(iv) this provides the following estimate

$$\int_{I\!R^s} |\Phi(h - Tx)| \, \mu(dh) = \int_{I\!R^s} |\Phi(h - Tx) - \Phi(0)| \, \mu(dh)$$

$$\leq \int_{I\!R^s} (\beta \|h - Tx\| + \gamma) \, \mu(dh)$$

$$\leq \beta \int_{I\!R^s} \|h\| \, \mu(dh) \; + \; \beta \|Tx\| + \gamma.$$

This implies that $Q_{I\!E}(x) \in I\!R$ for all $x \in I\!R^m$, and the proof is complete.

          q.e.d.

3.3   Continuity of the Probability Measure

Given a sequence $(M_n)_{n \in I\!N}$ of measurable sets in $I\!R^s$, the limes inferior $\liminf_{n \to \infty} M_n$ and the limes superior $\limsup_{n \to \infty} M_n$ are defined as the sets of all points belonging to all but a finite number of the M_n, and to infinitely many M_n, respectively. If μ is some probability measure on $I\!R^s$, then it holds

$$\mu\left(\liminf_{n \to \infty} M_n \right) \; \leq \; \liminf_{n \to \infty} \mu(M_n) \; \leq \; \limsup_{n \to \infty} \mu(M_n) \; \leq \; \mu\left(\limsup_{n \to \infty} M_n \right). \qquad (9)$$

This will be our main tool to deduce (semi-)continuity of the function $Q_{I\!P}$ from the properties of Φ. With $x \in I\!R^m$ we introduce the notation

$$M(x) := \{h \in I\!R^s \; : \; c^T x + \Phi(h - Tx) > \varphi_o\},$$

$$M_e(x) := \{h \in I\!R^s \; : \; c^T x + \Phi(h - Tx) = \varphi_o\},$$

$$M_d(x) := \{h \in I\!R^s \; : \; \Phi \text{ is discontinuous at } h - Tx\}.$$

Proposition 4. *The function $Q_{I\!P}$ is lower semicontinuous on $I\!R^m$. If, for some $x \in I\!R^m$, it holds $\mu\big(M_e(x) \cup M_d(x)\big) = 0$, then $Q_{I\!P}$ is continuous at x. The latter assumption is fulfilled for all $x \in I\!R^m$ if μ has a density.*

Proof: Let us first verify that for all $x \in I\!R^m$

$$M(x) \subseteq \liminf_{x_n \to x} M(x_n) \subseteq \limsup_{x_n \to x} M(x_n) \subseteq M(x) \cup M_e(x) \cup M_d(x). \quad (10)$$

Let $h \in M(x)$. The lower semicontinuity of Φ (Proposition 1(i)) yields

$$\liminf_{x_n \to x}(c^T x_n + \Phi(h - T x_n)) \geq c^T x + \Phi(h - Tx) > \varphi_o.$$

Therefore, there exists an $n_o \in I\!N$ such that $c^T x_n + \Phi(h - T x_n) > \varphi_o$ for all $n \geq n_o$, implying $h \in M(x_n)$ for all $n \geq n_o$. Hence, $M(x) \subseteq \liminf_{x_n \to x} M(x_n)$. Let $h \in \limsup_{x_n \to x} M(x_n) \setminus M(x)$. Then there exists an infinite subset $\tilde{I\!N}$ of $I\!N$ such that

$$c^T x_n + \Phi(h - T x_n) > \varphi_o \; \forall n \in \tilde{I\!N} \quad \text{and} \quad c^T x + \Phi(h - Tx) \leq \varphi_o.$$

Now two cases are posssible. First, Φ is continuous at $h - Tx$. Passing to the limit in the first inequality then yields that $c^T x + \Phi(h - Tx) \geq \varphi_o$, and $h \in M_e(x)$. Secondly, Φ is discontinuous at $h - Tx$. In other words, $h \in M_d(x)$, and (10) is established.

By (9) we have for all $x \in I\!R^m$

$$Q_{I\!P}(x) \; = \; \mu\big(M(x)\big) \leq \mu\big(\liminf_{x_n \to x} M(x_n)\big)$$

$$\leq \liminf_{x_n \to x} \mu\big(M(x_n)\big) \; = \; \liminf_{x_n \to x} Q_{I\!P}(x_n),$$

verifying the asserted lower semicontinuity. In case $\mu\big(M_e(x) \cup M_d(x)\big) = 0$ this argument extends:

$$Q_{I\!P}(x) \; = \; \mu\big(M(x)\big) \; = \; \mu\big(M(x) \cup M_e(x) \cup M_d(x)\big) \; \geq \; \mu\big(\limsup_{x_n \to x} M(x_n)\big)$$

$$\geq \; \limsup_{x_n \to x} \mu\big(M(x_n)\big) \; = \; \limsup_{x_n \to x} Q_{I\!P}(x_n),$$

and $Q_{I\!P}$ is continuous at x.

According to our discussion preceding Proposition 1 the set $M_d(x)$ is contained in a countable union of hyperplanes. In view of (8) the same is true for $M_e(x)$. Thus $M_e(x) \cup M_d(x)$ is contained in a set of Lebesgue measure zero, and $\mu\big(M_e(x) \cup M_d(x)\big) = 0$ by the absolute continuity of μ.   q.e.d.

3.4 Fatou's Lemma

For a sequence $(g_n)_{n \in I\!N}$ of measurable functions from $I\!R^s$ to $I\!R$ with an integrable minorant $g \leq g_n$, $\forall n \in I\!N$, Fatou's Lemma asserts

$$\int_{I\!R^s} \liminf_{n \to \infty} g_n(h) \, \mu(dh) \leq \liminf_{n \to \infty} \int_{I\!R^s} g_n(h) \, \mu(dh).$$

Together with the lower semicontinuity of Φ this will provide the lower semicontinuity of $Q_{I\!E}$.

Proposition 5. *The function Q_E is lower semicontinuous on $\mathbb{R}^m$ provided that $\int_{\mathbb{R}^s} \|h\| \, \mu(dh) < \infty$.*

Proof: Let $x \in \mathbb{R}^m$ and $x_n \to x$. We will apply Fatou's Lemma essentially to the functions $g_n(h) := \Phi(h - Tx_n)$. Denote $r := \max_{n \in \mathbb{N}} \|x_n\|$. Proposition 1(iv) and $\Phi(0) = 0$ then imply

$$\Phi(h - Tx_n) \geq \Phi(0) - |\Phi(h - Tx_n) - \Phi(0)|$$
$$\geq -\beta\|h - Tx_n\| - \gamma$$
$$\geq -\beta\|h\| - \beta r\|T\| - \gamma$$

yielding an integrable minorant g for the family of functions g_n. By Fatou's Lemma and the lower semicontinuity of Φ we have

$$Q_E(x) = \int_{\mathbb{R}^s} (c^T x + \Phi(h - Tx)) \, \mu(dh)$$
$$\leq \int_{\mathbb{R}^s} \liminf_{n \to \infty} (c^T x_n + \Phi(h - Tx_n)) \, \mu(dh)$$
$$\leq \liminf_{n \to \infty} \int_{\mathbb{R}^s} (c^T x_n + \Phi(h - Tx_n)) \, \mu(dh)$$
$$= \liminf_{n \to \infty} Q_E(x_n).$$

q.e.d.

3.5   Lebesgue's Dominated Convergence Theorem

Let g_n, g $(n \in \mathbb{N})$ be measurable functions from $\mathbb{R}^s$ to $\mathbb{R}$ fulfilling $\lim_{n \to \infty} g_n = g$, μ-almost surely.

If there exists an integrable function $\bar{g} \geq |g_n|$, $\forall n \in \mathbb{N}$, μ-almost surely, then Lebesgue's Dominated Convergence Theorem asserts that g_n, g $(n \in \mathbb{N})$ are integrable and that

$$\lim_{n \to \infty} \int_{\mathbb{R}^s} g_n(h) \, \mu(dh) = \int_{\mathbb{R}^s} g(h) \, \mu(dh).$$

This theorem will lead us to the continuity of Q_E.

Proposition 6. *If $\int_{\mathbb{R}^s} \|h\| \, \mu(dh) < \infty$ and $\mu(M_d(x)) = 0$ then Q_E is continuous at x. The latter assumption is fulfilled for all $x \in \mathbb{R}^m$ if μ has a density.*

Proof: Let $x \in \mathbb{R}^m$, $x_n \to x$, and $r := \max_{n \in \mathbb{N}} \|x_n\|$. Again we employ Proposition 1(iv) and $\Phi(0) = 0$, and we obtain

$$|\Phi(h - Tx_n)| = |\Phi(h - Tx_n) - \Phi(0)| \leq \beta\|h\| + \beta r\|T\| + \gamma$$

providing us with an integrable majorant. By $\mu(M_d(x)) = 0$, we have

$$\lim_{n \to \infty} (c^T x_n + \Phi(h - Tx_n)) = c^T x + \Phi(h - Tx) \quad \text{for } \mu\text{-almost all } h \in \mathbb{R}^s.$$

Now Lebesgue's Dominated Convergence Theorem completes the proof:

$$\lim_{n \to \infty} Q_E(x_n) = \lim_{n \to \infty} \int_{IR^s} (c^T x_n + \Phi(h - T x_n)) \, \mu(dh)$$

$$= \int_{IR^s} (c^T x + \Phi(h - T x)) \, \mu(dh) = Q_E(x).$$

q.e.d.

3.6 Convergence of Probability Measures – Rubin's Theorem

The dependence of the optimization problems (5) and (6) on the underlying probability measure, although seemingly a theoretical issue, has practical relevance in various respects. When building models like (5) and (6) the probability measure often enters in a subjective way or results from an approximation based on statistical data. Moreover, the integrals in (5) and (6) are typically multivariate, and their integrands are given only implicitly. This poses insurmountable numerical difficulties if the probability distribution is continuous. A possible remedy is approximation via discrete distributions. All this motivates considerations on whether "small" perturbations of the underlying probability measure result in "small" perturbations of optimal values and optimal solutions to (5) and (6). The mathematical machinery for addressing these issues is provided by the stability analysis of stochastic programs (for surveys see [11,22]).

A first and crucial step towards stability analysis is to study Q_E and Q_P as functions jointly in x and μ. Again the value function Φ will provide some valuable insights. Let us consider Q_E and Q_P as functions on $IR^m \times \mathcal{P}(IR^s)$ where $\mathcal{P}(IR^s)$ denotes the set of all (Borel) probability measures on IR^s. As an essential prerequisite some convergence notion is needed on $\mathcal{P}(IR^s)$. Here, weak convergence of probability measures has proven both sufficiently general to cover relevant applications and sufficiently specific to enable substantial statements. A sequence $(\mu_n)_{n \in IN}$ in $\mathcal{P}(IR^s)$ is said to converge weakly to $\mu \in \mathcal{P}(IR^s)$, written $\mu_n \overset{w}{\longrightarrow} \mu$, if for any bounded continuous function $g : IR^s \to IR$ we have

$$\int_{IR^s} g(\xi) \mu_n(d\xi) \to \int_{IR^s} g(\xi) \mu(d\xi) \quad \text{as} \quad n \to \infty. \tag{11}$$

A basic reference for weak convergence of probability measures is Billingsley's book [4].

We are heading for sufficient conditions for the continuity of Q_E and Q_P jointly in x and μ. Beside properties of the value function, a theorem on weak convergence of image measures attributed in [4] to Rubin will turn out most useful. This theorem says: Let g_n, g ($n \in IN$) be measurable functions from IR^s to IR and denote $E := \{h \in IR^s : \exists h_n \to h \text{ such that } g_n(h_n) \not\to g(h)\}$. If $\mu_n \overset{w}{\longrightarrow} \mu$ and $\mu(E) = 0$, then $\mu_n \circ g_n^{-1} \overset{w}{\longrightarrow} \mu \circ g^{-1}$.

Proposition 7. *Let $\mu \in \mathcal{P}(IR^s)$ be such that $\mu(M_e(x) \cup M_d(x)) = 0$. Then the function $Q_P : IR^m \times \mathcal{P}(IR^s) \longrightarrow IR$ is continuous at (x, μ).*

Proof: To prove (i), let $x_n \longrightarrow x$ (in $\mathbb{R}^m$) and $\mu_n \xrightarrow{w} \mu$ (in $\mathcal{P}(\mathbb{R}^s)$) be arbitrary sequences. By $\chi_n, \chi : \mathbb{R}^s \longrightarrow \{0, 1\}$ we denote the indicator functions of the sets $M(x_n), M(x), n \in \mathbb{N}$. With these functions we consider the exceptional set E from above:

$$E := \{h \in \mathbb{R}^s : \exists h_n \to h \text{ such that } \chi_n(h_n) \not\to \chi(h)\}.$$

To see that $E \subseteq M_e(x) \cup M_d(x)$, assume that $h \in \big(M_e(x) \cup M_d(x)\big)^c = \big(M_e(x)\big)^c \cap \big(M_d(x)\big)^c$ where the superscript c denotes the set-theoretic complement. Then Φ is continuous at $h - Tx$, and either $c^T x + \Phi(h - Tx) > \varphi_o$ or $c^T x + \Phi(h - Tx) < \varphi_o$. Thus, for any sequence $h_n \to h$ there is an $n_o \in \mathbb{N}$ such that, for all $n \geq n_o$, either $c^T x_n + \Phi(h_n - Tx_n) > \varphi_o$ or $c^T x_n + \Phi(h_n - Tx_n) < \varphi_o$. Hence, $\chi_n(h_n) \to \chi(h)$ as $h_n \to h$, implying $h \in E^c$.

In view of $E \subseteq M_e(x) \cup M_d(x)$ and $\mu\big(M_e(x) \cup M_d(x)\big) = 0$ we obtain that $\mu(E) = 0$. Rubin's Theorem now yields $\mu_n \circ \chi_n^{-1} \xrightarrow{w} \mu \circ \chi^{-1}$.

Since $\mu_n \circ \chi_n^{-1}, \mu \circ \chi^{-1}, n \in \mathbb{N}$ are probability measures on $\{0, 1\}$, their weak convergence particularly implies that

$$\mu_n \circ \chi_n^{-1}(\{1\}) \longrightarrow \mu \circ \chi^{-1}(\{1\}).$$

This is nothing but $\mu_n\big(M(x_n)\big) \longrightarrow \mu\big(M(x)\big)$ or $Q_{\mathbb{P}}(x_n, \mu_n) \longrightarrow Q_{\mathbb{P}}(x, \mu)$.

q.e.d.

Proposition 8. *Fix arbitrary $p > 1$ and $K > 0$, and denote $\Delta_{p,K}(\mathbb{R}^s) := \{\nu \in \mathcal{P}(\mathbb{R}^s) : \int_{\mathbb{R}^s} \|h\|^p \nu(dh) \leq K\}$. Let $\mu \in \Delta_{p,K}(\mathbb{R}^s)$ be such that $\mu\big(M_d(x)\big) = 0$. Then the function $Q_{\mathbb{E}} : \mathbb{R}^m \times \Delta_{p,K}(\mathbb{R}^s) \longrightarrow \mathbb{R}$ is continuous at (x, μ).*

Proof: Let $x_n \longrightarrow x$ in $\mathbb{R}^m$ and $\mu_n \xrightarrow{w} \mu$ in $\Delta_{p,K}(\mathbb{R}^s)$. Introduce measurable functions $g_n, n \in \mathbb{N}$, and g by $g_n(h) := \Phi(h - Tx_n)$ and $g(h) := \Phi(h - Tx)$. For the corresponding exceptional set E a simple continuity argument provides $M_d(x)^c \subseteq E^c$ or, equivalently, $E \subseteq M_d(x)$. Hence, $\mu(E) = 0$, and Rubin's Theorem yields

$$\mu_n \circ g_n^{-1} \xrightarrow{w} \mu \circ g^{-1}. \tag{12}$$

To prove the assertion it is sufficient to show that

$$\lim_{n \to \infty} \int_{\mathbb{R}^s} g_n(h) \, \mu_n(dh) = \int_{\mathbb{R}^s} g(h) \, \mu(dh).$$

Changing variables yields the equivalent statement

$$\lim_{n \to \infty} \int_{\mathbb{R}} t \, \mu_n \circ g_n^{-1}(dt) = \int_{\mathbb{R}} t \, \mu \circ g^{-1}(dt).$$

For fixed $a \in \mathbb{R}_+$, consider the truncation $\kappa_a : \mathbb{R} \to \mathbb{R}$ with

$$\kappa_a(t) := \begin{cases} t & , \ |t| < a \\ 0 & , \ |t| \geq a. \end{cases}$$

Now

$$\left| \int_{\mathbb{R}} t \ \mu_n \circ g_n^{-1}(dt) - \int_{\mathbb{R}} t \ \mu \circ g^{-1}(dt) \right|$$

$$\leq \left| \int_{\mathbb{R}} (t - \kappa_a(t)) \ \mu_n \circ g_n^{-1}(dt) \right|$$

$$+ \left| \int_{\mathbb{R}} \kappa_a(t) \ \mu_n \circ g_n^{-1}(dt) - \int_{\mathbb{R}} \kappa_a(t) \ \mu \circ g^{-1}(dt) \right|$$

$$+ \left| \int_{\mathbb{R}} (\kappa_a(t) - t) \ \mu \circ g^{-1}(dt) \right|. \tag{13}$$

The proof is completed by showing that, for a given $\varepsilon > 0$, each of the three expressions on the right becomes less than $\varepsilon/3$ provided that n and a are sufficiently large.

For the first expression we obtain

$$\left| \int_{\mathbb{R}} (t - \kappa_a(t)) \ \mu_n \circ g_n^{-1}(dt) \right| \leq \int_{\{t: |t| \geq a\}} |t| \ \mu_n \circ g_n^{-1}(dt)$$

$$= \int_{\{h: |g_n(h)| \geq a\}} |g_n(h)| \ \mu_n(dh). \tag{14}$$

Since $p > 1$,

$$\int_{\mathbb{R}^s} |g_n(h)|^p \ \mu_n(dh) \geq \int_{\{h: |g_n(h)| \geq a\}} |g_n(h)| \cdot |g_n(h)|^{p-1} \ \mu_n(dh)$$

$$\geq a^{p-1} \int_{\{h: |g_n(h)| \geq a\}} |g_n(h)| \ \mu_n(dh). \tag{15}$$

Therefore, the estimate in (14) can be continued by

$$\leq a^{1-p} \int_{\mathbb{R}^s} |g_n(h)|^p \ \mu_n(dh). \tag{16}$$

Proposition 1(iv) and $g_n(0) = 0$ imply

$$|g_n(h)|^p \leq (\beta\|h\| + \beta\|x_n\| \cdot \|T\| + \gamma)^p.$$

Since $(x_n)_{n \in \mathbb{N}}$ is bounded and all μ_n belong to $\Delta_{p,K}(\mathbb{R}^s)$, there exists a positive constant c such that

$$\int_{\mathbb{R}^s} |g_n(h)|^p \ \mu_n(dh) \leq c \quad \text{for all } n \in \mathbb{N}.$$

Hence, (16) can be estimated above by c/a^{p-1} which becomes less than $\varepsilon/3$ if a is sufficiently large.

We now turn to the second expression in (13).

Since every probability measure on the real line has at most countably many atoms, we obtain that $\mu \circ g^{-1}(\{t : |t| = a\}) = 0$ for (Lebesgue-)almost all $a \in \mathbb{R}$.

Therefore, κ_a is a measurable function whose set of discontinuity points D_{κ_a} has $\mu \circ g^{-1}$-measure zero for almost all $a \in \mathbb{R}$. We apply Rubin's Theorem to the weakly convergent sequence $\mu_n \circ g_n^{-1} \xrightarrow{w} \mu \circ g^{-1}$, cf. (12), and the identical sequence of functions κ_a. The role of the exceptional set then is taken by D_{κ_a}, and Rubin's Theorem is working due to $\mu \circ g^{-1}(D_{\kappa_a}) = 0$. This yields the conclusion

$$\mu_n \circ g_n^{-1} \circ \kappa_a^{-1} \xrightarrow{w} \mu \circ g^{-1} \circ \kappa_a^{-1} \qquad \text{for almost all } a \in \mathbb{R}. \tag{17}$$

Consider the bounded continuous function $\eta : \mathbb{R} \to \mathbb{R}$ given by

$$\eta(t') := \begin{cases} -a & , \quad t' \leq -a \\ t' & , \quad -a \leq t' \leq a \\ a & , \quad t' \geq a. \end{cases}$$

By the weak convergence in (17), we obtain

$$\int_{\mathbb{R}} \eta(t') \, \mu_n \circ g_n^{-1} \circ \kappa_a^{-1}(dt') \;\to\; \int_{\mathbb{R}} \eta(t') \, \mu \circ g^{-1} \circ \kappa_a^{-1}(dt') \qquad \text{as } n \to \infty. \tag{18}$$

Changing variables provides

$$\int_{\mathbb{R}} \eta(t') \, \mu_n \circ g_n^{-1} \circ \kappa_a^{-1}(dt') = \int_{\kappa_a^{-1}(\mathbb{R})} \eta(\kappa_a(t)) \, \mu_n \circ g_n^{-1}(dt)$$

$$= \int_{\mathbb{R}} \kappa_a(t) \, \mu_n \circ g_n^{-1}(dt).$$

Analogously,

$$\int_{\mathbb{R}} \eta(t') \, \mu \circ g^{-1} \circ \kappa_a^{-1}(dt') = \int_{\mathbb{R}} \kappa_a(t) \, \mu \circ g^{-1}(dt).$$

The above identities together with (18) confirm that the second expression on the right-hand side of (13) becomes arbitrarily small for sufficiently large n and almost all sufficiently large a.

Let us finally turn to the third expression in (13). Analogously to (14), (15), and (16) we obtain

$$\left| \int_{\mathbb{R}} (\kappa_a(t) - t) \, \mu \circ g^{-1}(dt) \right| \leq a^{1-p} \int_{\mathbb{R}^s} |g(h)|^p \, \mu(dh).$$

The integral $\int_{\mathbb{R}^s} |g(h)|^p \, \mu(dh)$ is finite due to Proposition 1(iv) and the fact that $\int_{\mathbb{R}^s} \|h\|^p \, \mu(dh) \leq K$. Hence, the third expression in (13) becomes less than $\varepsilon/3$ if a is large enough.   q.e.d.

4   Outlook

The previous sections have shown how the mixed-integer value function serves as a point of departure for understanding the basic structure of stochastic integer programs. Let us finally have a look at some further developments whose detailed coverage is beyond the scope of the present paper.

Quantitative Statements. The continuity results of Section 3 are all qualitative by nature. Lipschitz continuity of $Q_{I\!E}$ and $Q_{I\!P}$ as functions in x has been studied in [19,20,24]. To quantify the continuity of $Q_{I\!E}$ and $Q_{I\!P}$ as functions of the underlying probability measure a proper metric in the space of probability measures has to be identified. Here "proper" means that the metric should allow an estimation of function value distances at all and that it should metrize important modes of convergence such as weak convergence of probability measures. For the function $Q_{I\!E}$ a first proposal along these lines was made in [21]. The mentioned quantitative studies require as input refined statements about Φ such as parts (ii) and (iii) of Proposition 1.

Stability. As already mentioned in Subsection 3.6, perturbation and approximation of the underlying probability measure arise quite naturally in stochastic programming. The stability analysis of stochastic programs then provides justification for replacing unknown probability measures by statistical estimates or for turning numerically intractable multivariate integrals into manageable finite sums by approximating continuous distributions via discrete ones. Typical stability results assert that optimal values and optimal solutions are (semi-)continuous (multi-)functions of the underlying probability measure, [1,20,21]. These results are obtained by putting general techniques from parametric optimization into perspective with stochastic programming, [22]. This leads to studying the joint dependence of relevant integral functionals on both the decision variable and the probability measure. For the latter, Propositions 7 and 8 provide paradigmatic examples.

Algorithms. Methods for solving the optimization problems (5) and (6), almost exclusively, rest on the assumption that the probability measures underlying the models are discrete. This does not provide a serious restriction, since, on the one hand, in many practical situations the uncertain data is available via discrete observations, only. On the other hand, the above mentioned stability results justify approximation via discrete measures should the precise model involve a continuous probability distribution. With discrete probability measures, the problems (5) and (6) can be rewritten as large-scale, block-structured, mixed-integer linear programs. Decomposition then becomes the algorithmic method of choice, but the presence of integer variables poses a number of open problems. Some first attempts were made in [9,24], see also the survey [13].

References

1. Artstein, Z.; Wets, R.J-B: Stability results for stochastic programs and sensors, allowing for discontinuous objective functions, SIAM Journal on Optimization 4 (1994), 537–550.
2. Bank, B.; Guddat, J.; Klatte, D.; Kummer, B.; Tammer, K.: Non-linear Parametric Optimization, Akademie-Verlag, Berlin, 1982.

3. Bank, B.; Mandel, R.: Parametric Integer Optimization, Akademie-Verlag, Berlin 1988.
4. Billingsley, P.: Convergence of Probability Measures, Wiley, New York, 1968.
5. Billingsley, P.: Probability and Measure, Wiley, New York, 1986.
6. Birge, J.R.; Louveaux, F.: Introduction to Stochastic Programming, Springer, New York, 1997.
7. Blair, C.E.; Jeroslow, R.G.: The value function of a mixed integer program: I, Discrete Mathematics 19 (1977), 121–138.
8. Bonnans, J.F.; Shapiro, A.: Perturbation Analysis of Optimization Problems, Springer-Verlag, New York, 2000.
9. Carøe, C.C.; Schultz, R.: Dual decomposition in stochastic integer programming, Operations Research Letters 24 (1999), 37–45.
10. Dudley, R.M.: Real Analysis and Probability, Wadsworth & Brooks/Cole, Pacific Grove, California 1989.
11. Dupačová, J.: Stochastic programming with incomplete information: a survey of results on postoptimization and sensitivity analysis, Optimization 18 (1987), 507–532.
12. Kall, P.; Wallace, S.W.: Stochastic Programming, Wiley, Chichester, 1994.
13. Klein Haneveld, W.K.; van der Vlerk, M.H.: Stochastic integer programming: General models and algorithms, Annals of Opeations Research 85 (1999), 39–57.
14. Nemhauser, G.L.; Wolsey, L.A.: Integer and Combinatorial Optimization, Wiley, New York 1988.
15. Ogryczak, W.; Ruszczyński, A.: From stochastic dominance to mean-risk models: Semideviations as risk measures, European Journal of Operational Research 116 (1999), 33–50.
16. Ogryczak, W.; Ruszczyński, A.: Dual stochastic dominance and related mean-risk models, Rutcor Research Report 10–2001, Rutgers Center for Operations Research, Piscataway, 2001.
17. Prékopa, A.: Stochastic Programming, Kluwer, Dordrecht, 1995.
18. Rudin, W.: Real and Complex Analysis, McGraw-Hill, New York, 1974.
19. Schultz, R.: Continuity properties of expectation functions in stochastic integer programming, Mathematics of Operations Research 18 (1993), 578–589.
20. Schultz, R.: On structure and stability in stochastic programs with random technology matrix and complete integer recourse, Mathematical Programming 70 (1995), 73–89.
21. Schultz, R.: Rates of convergence in stochastic programs with complete integer recourse, SIAM Journal on Optimization 6 (1996), 1138–1152.
22. Schultz, R.: Some aspects of stability in stochastic programming, Annals of Operations Research 100 (2000), 55–84.
23. Schultz, R.: Probability objectives in stochastic programs with recourse, Preprint 506–2001, Institute of Mathematics, Gerhard-Mercator University Duisburg, 2001.
24. Tiedemann, S.: Probability Functionals and Risk Aversion in Stochastic Integer Programming, Diploma Thesis, Department of Mathematics, Gerhard-Mercator University Duisburg, 2001.

Exact Algorithms for NP-Hard Problems: A Survey

Gerhard J. Woeginger

Department of Mathematics
University of Twente, P.O. Box 217
7500 AE Enschede, The Netherlands

Abstract. We discuss fast exponential time solutions for NP-complete problems. We survey known results and approaches, we provide pointers to the literature, and we discuss several open problems in this area. The list of discussed NP-complete problems includes the travelling salesman problem, scheduling under precedence constraints, satisfiability, knapsack, graph coloring, independent sets in graphs, bandwidth of a graph, and many more.

1 Introduction

Every NP-complete problem can be solved by exhaustive search. Unfortunately, when the size of the instances grows the running time for exhaustive search soon becomes forbiddingly large, even for instances of fairly small size. For some problems it is possible to design algorithms that are significantly faster than exhaustive search, though still not polynomial time. This survey deals with such fast, super-polynomial time algorithms that solve NP-complete problems to optimality. In recent years there has been growing interest in the design and analysis of such super-polynomial time algorithms. This interest has many causes.

- It is now commonly believed that P≠NP, and that super-polynomial time algorithms are the best we can hope for when we are dealing with an NP-complete problem. There is a handful of isolated results scattered across the literature, but we are far from developing a general theory. In fact, we have not even started a systematic investigation of the worst case behavior of such super-polynomial time algorithms.
- Some NP-complete problems have better and faster exact algorithms than others. There is a wide variation in the worst case complexities of known exact (super-polynomial time) algorithms. Classical complexity theory can not explain these differences. Do there exist any relationships among the worst case behaviors of various problems? Is progress on the different problems connected? Can we somehow classify NP-complete problems to see how close we are to the best possible algorithms?
- With the increased speed of modern computers, large instances of NP-complete problems can be solved effectively. For example it is nowadays routine to solve travelling salesman (TSP) instances with up to 2000 cities.

M. Jünger et al. (Eds.): Combinatorial Optimization (Edmonds Festschrift), LNCS 2570, pp. 185–207, 2003.

And if the data is nicely structured, then instances with up to 13000 cities can be handled in practice (Applegate, Bixby, Chvátal & Cook [2]). There is a huge gap between the empirical results from testing implementations and the known theoretical results on exact algorithms.

- Fast algorithms with exponential running times may actually lead to practical algorithms, at least for moderate instance sizes. For small instances, an algorithm with an exponential time complexity of $O(1.01^n)$ should usually run much faster than an algorithm with a polynomial time complexity of $O(n^4)$.

In this article we survey known results and approaches to the worst case analysis of exact algorithms for NP-hard problems, and we provide pointers to the literature. Throughout the survey, we will also formulate many exercises and open problems. Open problems refer to unsolved research problems, while exercises pose smaller questions and puzzles that should be fairly easy to solve.

Organization of this survey. Section 2 collects some technical preliminaries and some basic definitions that will be used in this article. Sections 3–6 introduce and explain the four main techniques for designing fast exact algorithms: Section 3 deals with dynamic programming across the subsets, Section 4 discusses pruning of search trees, Section 5 illustrates the power of preprocessing the data, and Section 6 considers approaches based on local search. Section 7 discusses methods for proving *negative* results on the worst case behavior of exact algorithms. Section 8 gives some concluding remarks.

2   Technical Preliminaries

How do we measure the quality of an exact algorithm for an NP-hard problem? Exact algorithms for NP-complete problems are sometimes hard to compare, since their analysis is done in terms of different parameters. For instance, for an optimization problem on graphs the analysis could be done in terms of the number n of vertices, or possibly in the number m of edges. Since the standard reductions between NP-complete problems may increase the instance sizes, many questions in computational complexity theory depend delicately on the choice of parameters. The right approach seems to be to include an explicit complexity parameter in the problem specification (Impagliazzo, Paturi & Zane [21]). Recall that the decision version of every problem in NP can be formulated in the following way:

Given x, decide whether there exists y so that $|y| \leq m(x)$ and $R(x, y)$.

Here x is an instance of the problem; y is a short YES-certificate for this instance; $R(x, y)$ is a polynomial time decidable relation that verifies certificate y for instance x; and $m(x)$ is a polynomial time computable and polynomially bounded *complexity parameter* that bounds the length of the certificate y. A trivial exact algorithm for solving x would be to enumerate all possible strings with lengths up to $m(x)$, and to check whether any of them yields a YES-certificate. Up to polynomial factors that depend on the evaluation time of $R(x, y)$, this would yield a running time of $2^{m(x)}$. The first goal in exact algorithms always is to break the triviality barrier, and to improve on the time complexity of this trivial enumerative algorithm.

Throughout this survey, we will measure the running times of algorithms *only* with respect to the complexity parameter $m(x)$. We will use a modified big-Oh notation that suppresses all other (polynomially bounded) terms that depend on the instance x and the relation $R(x, y)$. We write $O^*(T(m(x)))$ for a time complexity of the form $O(T(m(x)) \cdot \text{poly}(|x|))$. This modification may be justified by the exponential growth of $T(m(x))$. Note that for instance for simple graphs with $m(x) = n$ vertices and m edges, the running time $1.7344^n \cdot n^2 m^5$ is sandwiched between the running times 1.7344^n and 1.7345^n.

We stress, however, the fact that the complexity parameter $m(x)$ in general is not unique, and that it heavily depends on the representation of the input. For an input in the form of an undirected graph, for instance, the complexity parameter might be the number n of vertices or the number m of edges.

Time complexities and complexity classes. Consider a problem in NP as defined above, with instances x and with complexity parameter $m(x)$. An algorithm for this problem has *sub-exponential* time complexity, if the running time depends polynomially on $|x|$ and if the logarithm of the running time depends sub-linearly on $m(x)$. For instance, a running time of $|x|^5 \cdot 2^{\sqrt{m(x)}}$ would be sub-exponential. A problem in NP is contained in the complexity class SUBEXP (the class of SUB-EXPonentially solvable problems) if for every fixed $\varepsilon > 0$, it can be solved in $\text{poly}(|x|) \cdot 2^{\varepsilon \cdot m(x)}$ time.

The complexity class SNP (the class Strict NP) was introduced by Papadimitriou & Yannakakis [32] for studying the approximability of optimization problems. SNP constitutes a subclass of NP, and it contains all problems that can be formulated in a certain way by a logical formula that starts with a series of second order existential quantifiers, followed by a series of first order universal quantifiers, followed by a first-order quantifier-free formula (a Boolean combination of input and quantifier relations applied to the quantified element variables). In this survey, the class SNP will only show up in Section 7. As far as this survey is concerned, all we need to know about SNP is that it is a fairly broad complexity class that contains many of the natural combinatorial optimization problems.

Downey & Fellows [7] introduced parameterized complexity theory for investigating the complexity of problems that involve a parameter. This parameter may for instance be the treewidth or the genus of an underlying graph, or an upper bound on the objective value, or in our case the complexity parameter $m(x)$. A whole theory has evolved around such parameterizations, and this has lead to the so-called W-hierarchy, an infinite hierarchy of complexity classes:

$$FPT \subseteq W[1] \subseteq W[2] \subseteq \cdots \subseteq W[k] \subseteq \cdots \subseteq W[P].$$

We refer the reader to Downey & Fellows [8] for the exact definitions of all these classes. It is commonly believed that all W-classes are pairwise distinct, and that hence all displayed inclusions are strict.

Some classes of optimization problems. Let us briefly discuss some basic classes of optimization problems that contain many classical problems: the class of subset problems, the class of permutation problems, and the class of partition problems. In a *subset problem*, every feasible solution can be specified as a subset of an underlying ground set. For instance, fixing a truth-assignment in the satisfiability problem corresponds to selecting a subset of TRUE variables. In the independent set problem, every subset of the vertex set is a solution candidate. In a *permutation problem*, every feasible solution can be specified as a total ordering of an underlying ground set. For instance, in the TSP every tour corresponds to a permutation of the cities. In single machine scheduling problems, feasible schedules are often specified as permutations of the jobs. In a *partition problem*, every feasible solution can be specified as a partition of an underlying ground set. For instance, a graph coloring is a partition of the vertex set into color classes. In parallel machine scheduling problems, feasible schedules are often specified by partitioning the job set and assigning every part to another machine.

As we observed above, all NP-complete problems possess trivial algorithms that simply enumerate and check all feasible solutions. For a ground set with cardinality n, subset problems can be trivially solved in $O^*(2^n)$ time, permutation problems can be trivially solved in $O^*(n!)$ time, and partition problems are trivial to solve in $O^*(c^{n \log n})$ time; here $c > 1$ denotes a constant that does not depend on the instance. These time complexities form the triviality barriers for the corresponding classes of optimization problems.

More technical remarks. All optimization problems considered in this survey are known to be NP-complete. We refer the reader to the book [14] by Garey & Johnson for (references to) the NP-completeness proofs. We denote the base two logarithm of a real number z by $\log(z)$.

3   Technique: Dynamic Programming across the Subsets

A standard approach for getting fast exact algorithms for NP-complete problems is to do dynamic programming across the subsets. For every 'interesting' subset of the ground set, there is a polynomial number of corresponding states in the state space of the dynamic program. In the cases where all these corresponding states can be computed in reasonable time, this approach usually yields a time complexity of $O^*(2^n)$. We will illustrate these benefits of dynamic programming by developing algorithms for the travelling salesman problem and for total completion time scheduling on a single machine under precedence constraints. Sometimes, the number of 'interesting' subsets is fairly small, and then an even better time complexity might be possible. This will be illustrated by discussing the graph 3-colorability problem.

The travelling salesman problem (TSP). A travelling salesman has to visit the cities 1 to n. He starts in city 1, runs through the remaining $n - 1$ cities in arbitrary order, and in the very end returns to his starting point in city 1. The distance from city i to city j is denoted by $d(i, j)$. The goal is to minimize the total travel length of the salesman. A trivial algorithm for the TSP checks all $O(n!)$ permutations.

We now sketch the exact TSP algorithm of Held & Karp [16] that is based on dynamic programming across the subsets. For every non-empty subset $S \subseteq \{2, \ldots, n\}$ and for every city $i \in S$, we denote by $\text{OPT}[S; i]$ the length of the shortest path that starts in city 1, then visits all cities in $S - \{i\}$ in arbitrary order, and finally stops in city i. Clearly, $\text{OPT}[\{i\}; i] = d(1, i)$ and

$$\text{OPT}[S; i] = \min \{\text{OPT}[S - \{i\}; j] + d(j, i) : j \in S - \{i\}\}.$$

By working through the subsets S in order of increasing cardinality, we can compute the value $\text{OPT}[S; i]$ in time proportional to $|S|$. The optimal travel length is given as the minimum value of $\text{OPT}[\{2, \ldots, n\}; j] + d(j, 1)$ over all j with $2 \leq j \leq n$. This yields an overall time complexity of $O(n^2 2^n)$ and hence $O^*(2^n)$.

This result was published in 1962, and from nowadays point of view almost looks trivial. Still, it yields the best time complexity that is known today.

Open problem 31 *Construct an exact algorithm for the travelling salesman problem with time complexity $O^*(c^n)$ for some $c < 2$. In fact, it even would be interesting to reach such a time complexity $O^*(c^n)$ with $c < 2$ for the closely related, but slightly simpler Hamiltonian cycle problem (given a graph G on n vertices, does it contain a spanning cycle).*

Hwang, Chang & Lee [19] describe a sub-exponential time $O(c^{\sqrt{n} \log n})$ exact algorithm with some constant $c > 1$ for the Euclidean TSP. The Euclidean TSP is a special case of the TSP where the cities are points in the Euclidean plane and where the distance between two cities is the Euclidean distance. The approach in [19] is heavily based on planar separator structures, and it cannot be carried over to the general TSP. The approach can be used to yield similar time bounds for various NP-complete geometric optimization problems, as the Euclidean p-center problem and the Euclidean p-median problem.

Total completion time scheduling under precedence constraints. There is a single machine, and there are n jobs $1, \ldots, n$ that are specified by their length p_j and their weight w_j ($j = 1, \ldots, n$). Precedence constraints are given by a partial order on the jobs; if job i precedes job j in the partial order (denoted by $i \to j$), then i must be processed to completion before j can begin its processing. All jobs are available at time 0. We only consider non-preemptive schedules, in which all p_j time units of job J_j must be scheduled consecutively. The goal is to schedule the jobs on the single machine such that all precedence constraints are obeyed and such that the total completion time $\sum_{j=1}^{n} w_j C_j$ is minimized; here C_j is the time at which job j is completed in the given schedule. A trivial algorithm checks all $O(n!)$ permutations of the jobs.

Dynamic programming across the subsets yields a time complexity of $O^*(2^n)$. A subset $S \subseteq \{1, \ldots, n\}$ of the jobs is called an *ideal*, if $j \in S$ and $i \to j$ always implies $i \in S$. In other words, for every job $j \in S$ the ideal S also contains all jobs that have to be processed before j. For an ideal S, we denote by FIRST(S) all jobs in S without predecessors, by LAST(S) all jobs in S without successors, and by $p(S) = \sum_{i \in S} p_i$ the total processing time of the jobs in S. For an ideal S, we denote by OPT[S] the smallest possible total completion time for the jobs in S. Clearly, for any $j \in$ FIRST($\{1, \ldots, n\}$) we have OPT[$\{j\}$] $= w_j p_j$. Moreover, for $|S| \geq 2$ we have

$$\text{OPT}[S] = \min\left\{\text{OPT}[S - \{j\}] + w_j \, p(S) : \; j \in \text{LAST}(S)\right\}.$$

This DP recurrence is justified by the observation that some job $j \in$ LAST(S) has to be processed last, and thus is completed at time $p(S)$. By working through the ideals S in order of increasing cardinality, we can compute all values OPT[S] in time proportional to $|S|$. The optimal objective value can be read from OPT[$\{1, \ldots, n\}$]. This yields an overall time complexity of $O^*(2^n)$.

Similar approaches yield $O^*(c^n)$ time exact algorithms for many other single machine scheduling problems.

Exercise 32 *Use dynamic programming across the subsets to get exact algorithms with time complexity $O^*(2^n)$ for the following two scheduling problems.*

(a) Minimizing the weighted number of late jobs. There are n jobs that are specified by a length p_j, a penalty w_j, and a due date d_j. If a job j is completed after its due date d_j, one has to pay a penalty p_j. The goal is to sequence the jobs on a single machine such that the total penalty for the late jobs is minimized.

(b) Minimizing the total tardiness. There are n jobs that are specified by a length p_j and a due date d_j. If a job j is completed at time C_j in some fixed schedule, then its tardiness is $T_j = \max\{0, C_j - d_j\}$. The goal is to sequence the jobs on a single machine such that the total tardiness of the jobs is minimized.

Exercise 33 *Total completion time scheduling under precedence constraints and job release dates. That is the problem that we have solved above, but with the additional restriction that every job j cannot be processed before its release r_j. As a consequence, their might be gaps in the middle of the schedule where the machine is idle.*

Use dynamic programming across the subsets to get an exact algorithm with time complexity $O^*(3^n)$ for this problem.

Graph coloring. Given a graph $G = (V, E)$ with n vertices, color the vertices with the smallest possible number of colors such that adjacent vertices never receive the same color. This smallest possible number is the *chromatic* number $\chi(G)$ of the graph. Every color class is a vertex set without induced edges; such a vertex set is called an *independent* set. An independent set is *maximal*, if none of its proper supersets is also independent. For any graph G, there exists a feasible coloring with $\chi(G)$ colors in which at least one color class is a maximal independent set. Moon & Moser [29] have shown that a graph with n vertices contains at most $3^{n/3} \approx 1.4422^n$ maximal independent sets. By considering a collection of $n/3$ independent triangles, we see that this bound is best possible. Paull & Unger [36] designed a procedure that generates all maximal independent sets in a graph in $O(n^2)$ time per generated set.

Based on the ideas introduced by Lawler [26], we present a dynamic program across the subsets with a time complexity of $O^*(2.4422^n)$. For a subset $S \subseteq V$ of the vertices, we denote by $G[S]$ the subgraph of G that is induced by the vertices in S, and we denote by $\text{OPT}[S]$ the chromatic number of $G[S]$. If S is empty, then clearly $\text{OPT}[S] = 0$. Moreover, for $S \neq \emptyset$ we have

$$\text{OPT}[S] = 1 + \min\{\text{OPT}[S - T] : T \text{ maximal indep. set in } G[S]\}.$$

We work through the sets S in order of increasing cardinality, such that when we are handling S, all its subsets have already been handled. Then the time needed to compute the value $\text{OPT}[S]$ is dominated by the time needed to generate all maximal independent subsets T of $G[S]$. By the above discussion, this can be done in $k^2 3^{k/3}$ time where k is the number of vertices in $G[S]$. This leads to an overall time complexity of

$$\sum_{k=0}^{n} \binom{n}{k} k^2 3^{k/3} \leq n^2 \sum_{k=0}^{n} \binom{n}{k} 3^{k/3} = n^2 (1 + 3^{1/3})^n.$$

Since $1 + 3^{1/3} \approx 2.4422$, this yields the claimed time complexity $O^*(2.4422^n)$. Very recently, Eppstein [11] managed to improve this time complexity to $O^*(2.4150^n)$ where $2.4150 \approx 4/3 + 3^{4/3}/4$. His improvement is based on carefully counting the *small* maximal independent sets in a graph.

Finally, we turn to the (much easier) special case of deciding whether $\chi(G) = 3$. Lawler [26] gives a simple $O^*(1.4422^n)$ algorithm: Generate all maximal independent sets S, and check whether their complement graph $G[V - S]$ is bipartite. Schiermeyer [42] describes a rather complicated modification of this idea that improves the time complexity to $O^*(1.415^n)$. The first major progress is due to Beigel & Eppstein [4] who get a running time of $O^*(1.3446^n)$ by applying the technique of pruning the search tree; see Section 4 of this survey. The current champion algorithm has a time complexity of $O^*(1.3289^n)$ and is due to Eppstein [10]. This algorithm combines pruning of the search tree with several tricks based on network flows and matching.

Exercise 34 *(Nielsen [30])*
Find an $O^*(1.7851^n)$ exact algorithm that decides for a graph on n vertices whether $\chi(G) = 4$. Hint: Generate all maximal independent sets of cardinality at least $n/4$ (why?), and use the algorithm from [10] to check their complement graphs.

Eppstein [10] also shows that for $n/4 \leq k \leq n/3$, a graph on n vertices contains at most $O(3^{4k-n}4^{n-3k})$ maximal independent sets. Apply this result to improve the time complexity for 4-coloring further to $O^*(1.7504^n)$.

Open problem 35 *Design fast algorithms for k-colorability where k is small, say for $k \leq 6$. Design faster exact algorithms for the general graph coloring problem. Can we reach running times around $O^*(2^n)$?*

4   Technique: Pruning the Search Tree

Every NP-complete problem can be solved by enumerating and checking all feasible solutions. An organized way for doing this is to (1) concentrate on some piece of the feasible solution, to (2) determine all the possible values this piece can take, and to (3) branch into several subcases according to these possible values. This naturally defines a search tree: Every branching in (3) corresponds to a branching of the search tree into subtrees. Sometimes, it can be argued that certain values for a certain piece can never lead to an optimal solution. In these cases we may simply ignore all these values, kill the corresponding subtrees, and speed-up the search procedure. Every Branch-and-Bound algorithm is based on this idea, and we can also get exact algorithms with good worst case behavior out of this idea. However, to get the worst case analysis through, we need a good mathematical understanding of the evolution of the search tree, and we need good estimates on the sizes of the killed subtrees and on the number and on the sizes of the surviving cases.

We will illustrate the technique of pruning the search tree by developping algorithms for the satisfiability problem, for the independent set problem in graphs, and for the bandwidth problem in graphs.

The satisfiability problem. Let $X = \{x_1, x_2, \ldots, x_n\}$ be a set of logical variables. A variable or a negated variable from X is called a *literal*. A *clause* over X is the disjunction of literals from X. A Boolean formula is in *conjunctive normal form (CNF)*, if it is the conjunction of clauses over X. A formula in CNF is in *k-CNF*, if all clauses contain at most k literals. A formula is *satisfiable*, if there is a truth assignment from X to $\{0, 1\}$ which assigns to each variable a Boolean value (0=false, 1=true) such that the entire formula evaluates to true. The k-satisfiability problem is the problem of deciding whether a formula F in k-CNF is satisfiable. It is well-known that 2-satisfiability is polynomially solvable, whereas k-satisfiability with $k \geq 3$ is NP-complete. A trivial algorithm checks all possible truth assignments in $O^*(2^n)$ time.

We will now describe an exact $O^*(1.8393^n)$ algorithm for 3-satisfiability that is based on the technique of pruning the search tree. Let F be a Boolean formula in 3-CNF with m clauses ($m \leq n^3$). The idea is to branch on one of the clauses c with three literals ℓ_1, ℓ_2, ℓ_3. Every satisfying truth assignment for F must fall into one of the following three classes:

(a) literal ℓ_1 is true;
(b) literal ℓ_1 is false, and literal ℓ_2 is true;
(c) literals ℓ_1 and ℓ_2 are false, and literal ℓ_3 is true.

We fix the values of the corresponding one, two, three variables appropriately, and we branch into three subtrees according to these cases (a), (b), and (c) with $n-1$, $n-2$, and $n-3$ unfixed variables, respectively. By doing this, we cut away the subtree where the literals ℓ_1, ℓ_2, ℓ_3 all are false. The formulas in the three subtrees are handled recursively. The stopping criterion is when we reach a formula in 2-CNF, which can be resolved in polynomial time. Denote by $T(n)$ the worst case time that this algorithm needs on a 3-CNF formula with n variables. Then

$$T(n) \ \leq \ T(n-1) + T(n-2) + T(n-3) \ + O(n+m).$$

Here the terms $T(n-1)$, $T(n-2)$, and $T(n-3)$ measure the time for solving the subcase with $n-1$, $n-2$, and $n-3$ unfixed variables, respectively. Standard calculations yield that $T(n)$ is within a polynomial factor of α^n where α is the largest real root of $\alpha^3 = \alpha^2 + \alpha + 1$. Since $\alpha \approx 1.8393$, this gives a time complexity of $O^*(1.8393^n)$.

In a milestone paper in this area, Monien & Speckenmeyer [28] improve the branching step of the above approach. They either detect a clause that can be handled without any branching, or they detect a clause for which the branching only creates formulas that contain one clause with at most $k-1$ literals. A careful analysis yields a time complexity of $O^*(\beta^n)$ for k-satisfiability, where β is the largest real root of $\beta = 2 - 1/\beta^{k-1}$. For 3-satisfiability, this time complexity is $O^*(1.6181^n)$. Schiermeyer [41] refines these ideas for 3-satisfiability even further, and performs a quantitative analysis of the number of 2-clauses in the resulting subtrees. This yields a time complexity of $O^*(1.5783^n)$. Kullmann [24, 25] writes half a book on the analysis of this approach, and gets time complexities of $O^*(1.5045^n)$ and $O^*(1.4963^n)$ for 3-satisfiability. The current champion algorithms for satisfiability are, however, not based on pruning the search tree, but on local search ideas; see Section 6 of this survey.

Exercise 41 *For a formula F in CNF, consider the following bipartite graph G_F: For every logical variable in X, there is a corresponding variable-vertex in G_F, and for every clause in F, there is a corresponding clause-vertex in G_F. There is an edge from a variable-vertex to a clause-vertex if and only if the corresponding variable is contained (in negated or un-negated form) in the corresponding clause. The planar satisfiability problem is the special case of the satisfiability problem that contains all instances with formulas F in CNF for which the graph G_F is planar.*

Design a sub-exponential time exact algorithm for the planar 3-satisfiability problem! Hint: Use the planar separator theorem of Lipton & Tarjan [27] to break the formula F into two smaller, independent pieces. Running times of roughly $O^*(c^{\sqrt{n}})$ are possible.

The independent set problem. Given a graph $G = (V, E)$ with n vertices, the goal is to find an independent set of maximum cardinality. An *independent* set $S \subseteq V$ is a set of vertices that does not induce any edges. Moon & Moser [29] have shown that a graph contains at most $3^{n/3} \approx 1.4422^n$ maximal (with respect to inclusion) independent sets. Hence the first goal is to beat the time complexity $O^*(1.4422^n)$.

We describe an exact $O^*(1.3803^n)$ algorithm for independent set that is based on the technique of pruning the search tree. Let G be a graph with m edges. The idea is to branch on a high-degree vertex: If all vertices have degree at most two, then the graph is a collection of cycles and paths. It is straightforward to determine a maximum independent set in such a graph. Otherwise, G contains a vertex v of degree $d \geq 3$; let $v_1, \ldots, v_d$ be the neighbors of v in G. Every independent set I for G must fall into one of the following two classes:

(a) I does not contain v.
(b) I does contain v; then I cannot contain any neighbor of v.

We dive into two subtrees. The first subtree deals with the graph that results from removing vertex v from G. The second subtree deals with the graph that results from removing v together with $v_1, \ldots, v_d$ from G. We recursively compute the maximum independent set in both subtrees, and update it to a solution for the original graph G. Denote by $T(n)$ the worst case time that this algorithm needs on a graph with n vertices. Then

$$T(n) \leq T(n-1) + T(n-4) + O(n+m).$$

Standard calculations yield that $T(n)$ is within a polynomial factor of γ^n where $\gamma \approx 1.3803$ is the largest real root of $\gamma^4 = \gamma^3 + 1$. This yields the time complexity $O^*(1.3803^n)$.

The first published paper that deals with exact algorithms for maximum independent set is Tarjan & Trojanowski [46]. They give an algorithm with running time $O^*(1.2599^n)$. This algorithm follows essentially the above approach, but performs a smarter (and pretty tedious) structural case analysis of the neighborhood around the high-degree vertex v. The algorithm of Jian [22] has a time complexity of $O^*(1.2346^n)$. Robson [38] further refines the approach. A combinatorial argument about connected regular graphs helps to get the running time down to $O^*(1.2108^n)$. Robson's algorithm uses exponential space. Beigel [3] presents another algorithm with a weaker time complexity of $O^*(1.2227^n)$, but polynomial space complexity. Robson [39] is currently working on a new algorithm which is supposed to run in time $O^*(1.1844^n)$. This new algorithm is based on a detailed computer generated subcase analysis where the number of subcases is in the tens of thousands.

Open problem 42 *(a) Construct an exact algorithm for the maximum independent set problem with time complexity $O^*(c^n)$ for some $c \leq 1.1$. If this really is doable, it will be very tedious to do.*

(b) Prove a lower bound on the time complexity of any exact algorithm for maximum independent set that is based on the technique of pruning the search tree and that makes its branching decision by solely considering the subgraphs around a fixed chosen vertex.

Exercise 43 *(a) Design an algorithm with time complexity $O^*(1.1602^n)$ for the restriction of the maximum independent set problem to graphs with maximum degree three! Warning: This is not an easy exercise. See Chen, Kanj & Jia [5] for a solution.*

(b) Design a sub-exponential time exact algorithm for the restriction of the maximum independent set problem to planar graphs! Hint: Use the planar separator theorem of Lipton & Tarjan [27].

Open problem 44 *An input to the Max-Cut problem consists of a graph $G = (V, E)$ on n vertices. The goal is to find a partition of V into two sets V_1 and V_2 that maximizes the number of edges between V_1 and V_2 in E.*

(a) Design an exact algorithm for the Max-Cut problem with time complexity $O^*(c^n)$ for some $c < 2$.

(b) Design an exact algorithm for the restriction of the Max-Cut problem to graphs with maximum degree three that has a time complexity $O^*(c^n)$ for some $c < 1.5$. Gramm & Niedermeier [15] state an algorithm with time complexity $O^*(1.5160^n)$.

The bandwidth problem. Given a graph $G = (V, E)$ with n vertices, a *linear arrangement* is a bijective numbering $f : V \rightarrow \{1, \ldots, n\}$ of the vertices from 1 to n (which can be viewed as a layout of the graph vertices on a line). In some fixed linear arrangement, the *stretch* of an edge $[u, v] \in E$ is the distance $|f(u) - f(v)|$ of its endpoints, and the *bandwidth* of the linear arrangement is the maximum stretch over all edges. In the bandwidth problem, the goal is to find a linear arrangement of minimum bandwidth for G. A trivial algorithm checks all possible linear arrangements in $O^*(n!)$ time.

We will sketch an exact $O^*(20^n)$ algorithm for the bandwidth problem that is based on the technique of pruning the search tree. This beautiful algorithm is due to Feige & Kilian [13]. The algorithm checks for every integer b with $1 \leq b \leq n$ in $O^*(20^n)$ time whether the bandwidth of the input graph G is less or equal to b. To simplify the presentation, we assume that both n and b are powers of two (and otherwise analogous but more messy arguments go through). Moreover, we assume that G is connected. The algorithm proceeds in two phases. In the first phase, it generates an initial piece of the search tree that branches into up to 5^n subtrees. In the second phase, each of these subtrees is handled in $O(4^n)$ time per subtree.

The goal of the first phase is to break the set of 'reasonable' linear arrangements into up to $n\,5^{n-1}$ subsets; in each of these subsets the approximate position of every single vertex is known. More precisely, we partition the interval $[1, n]$ into $2n/b$ segments of length $b/2$, and we will assign every vertex to one of these segments. We start with an arbitrary vertex $v \in V$, and we check all $2n/b$ possibilities for assigning v to some segment. Then we iteratively select a yet unassigned vertex x that has a neighbor y that has already been assigned to some segment. In any linear arrangement with bandwidth b, vertex x can not be placed more than two segments away from vertex y; hence, vertex x can only be assigned to five possible segments. There are $n - 1$ vertices to assign, and we end up with $O^*(5^n)$ assignments.

In the second phase, we check which of these $O^*(5^n)$ assignments can be extended to a linear arrangement of bandwidth at most b. All assignments are handled in the same way: If an assignment stretches some edge from a segment to another segment with at least two other segments in between, then this assignment can never lead to a linear arrangement with bandwidth b; therefore, we may kill such an assignment right away. If an edge goes from a segment to the same segment or to one of the adjacent segments, then it will have stretch at most b regardless of the exact positions of vertices in segments; therefore, such an edge may be removed. Hence, in the end we are only left with edges that either connect consecutive even numbered segments or consecutive odd numbered segments. The problem decomposes into two independent subproblems, one within the even numbered segments and one within the odd numbered segments.

All these subproblems now are solved recursively. We break every segment into two subsegments of equal length. We try all possibilities for assigning every vertex from every segment into the corresponding left and right subsegments. Some of these refined assignments can be killed right away since they overstretch some edge; other edges are automatically short, and hence can be removed. In any case, we end up with two independent subproblems (one within the right subsegments and one within the left subsegments) that both can be solved recursively. Denote by $T(k)$ the time needed for solving a subproblem with k vertices. Then

$$T(k) \leq 2^k \cdot (T(k/2) + T(k/2)).$$

Standard calculations yield that $T(k)$ is in $O(4^k)$. Therefore, in the second phase we check $O^*(5^n)$ assignments in $O^*(4^n)$ time per assignment. This yields an overall time complexity of $O^*(20^n)$. Feige & Kilian [13] do a more careful analysis and improve the time complexity below $O^*(10^n)$.

Open problem 45 *(Feige & Kilian [13])*
Does the bandwidth problem have considerably faster exact algorithms? For instance, can it be solved in $O^*(2^n)$ time?

Exercise 46 *In the minimum sum linear arrangement problem, the input is a graph $G = (V, E)$ with n vertices. The goal is to find a linear arrangement of G that minimizes the sum of the stretches of all edges. Design an exact algorithm with time complexity $O^*(2^n)$ for this problem. Hint: Do not use the technique of pruning the search tree.*

5   Technique: Preprocessing the Data

Preprocessing is an initial phase of computation, where one analyzes and restructures the given data, such that later on certain queries to the data can be answered quickly. By preprocessing an exponentially large data set or part of this data in an appropriate way, we may sometimes gain an exponentially large factor in the running time. In this section we will use the technique of preprocessing the data to get fast algorithms for the subset sum problem and for the binary knapsack problem. We start this section by discussing two very simple, polynomially solvable toy problems where preprocessing helps a lot.

In the first toy problem, we are given two integer sequences $x_1, \ldots, x_k$ and $y_1, \ldots, y_k$ and an integer S. We want to decide whether there exist an x_i and a y_j that sum up to S. A trivial approach would be to check all possible pairs in $O(k^2)$ overall time. A better approach is to first preprocess the data and to sort the x_i in $O(k \log k)$ time. After that, we may repeatedly use bisection search in this sorted array, and search for the k values $S - y_j$ in $O(\log k)$ time per value. The overall time complexity becomes $O(k \log k)$, and we save a factor of $k / \log k$. By applying the same preprocessing, we can also decide in $O(k \log k)$ time, whether the sequences $\langle x_i \rangle$ and $\langle y_j \rangle$ are disjoint, or whether every value x_i also occurs in the sequence $\langle y_j \rangle$.

In the second toy problem, we are given k points (x_i, y_i) in two-dimensional space, together with the n numbers $z_1, \ldots, z_k$, and a number W. The goal is to determine for every z_j the largest value y_i, subject to the condition that $x_i + z_j \leq W$. The trivial solution needs $O(k^2)$ time, and by applying preprocessing this can be brought down to $O(k \log k)$: If there are two points (x_i, y_i) and (x_j, y_j) with $x_i \leq x_j$ and $y_i \geq y_j$, then the point (x_j, y_j) may be disregarded since it is always dominated by (x_i, y_i). The subset of non-dominated points can be computed in $O(k \log k)$ time by standard methods from computational geometry. We sort the non-dominated points by increasing x-coordinates and store this sequence in an array. This completes the preprocessing. To handle a value z_j, we simply search in $O(\log k)$ time through the sorted array for the largest value x_i less or equal to $W - z_j$.

In both toy problems preprocessing improved the time complexity from $O(k^2)$ to $O(k \log k)$. Of course, when dealing with exponential time algorithms an improvement by a factor of $k / \log k$ is not impressive at all. The right intuition is to think of k as roughly $2^{n/2}$. Then preprocessing the data yields a speedup from $k^2 = 2^n$ to $k \log k = n 2^{n/2}$, and such a speedup of $2^{n/2}$ indeed *is* impressive!

The subset sum problem. In this problem, the input consists of positive integers $a_1, \ldots, a_n$ and S. The question is whether there exists a subset of the a_i that sums up to S. The subset sum problem belongs to the class of subset problems, and can be solved (trivially) in $O^*(2^n)$ time. By splitting the problem into two halves and by preprocessing the first half, the time complexity can be brought down to $O^*(\sqrt{2}^n) \approx O^*(1.4145^n)$.

Let X denote the set of all integers of the form $\sum_{i \in I} a_i$ with $I \subseteq \{1, \ldots, \lfloor n/2 \rfloor\}$, and let Y denote the set of all integers of the form $\sum_{i \in I} a_i$ with $I \subseteq \{\lfloor n/2 \rfloor + 1, \ldots, n\}$. Note that $0 \in X$ and $0 \in Y$. It is straightforward to compute X and Y in $O^*(2^{n/2})$ time by complete enumeration. The subset sum instance has a solution if and only if there exists an $x_i \in X$ and a $y_j \in Y$ with $x_i + y_j = S$. But now we are back at our first toy problem that we discussed at the beginning of this section! By preprocessing X and by searching for all $S - y_j$ in the sorted structure, we can solve this problem in $O(n2^{n/2})$ time. This yields an overall time of $O^*(2^{n/2})$.

Exercise 51 *An input to the Exact-Hitting-Set problem consists of a ground set X with n elements, and a collection S of subsets over X. The question is whether there exists a subset $Y \subseteq X$ such that $|Y \cap T| = 1$ for all $T \in S$.*

Use the technique of preprocessing the data to get an exact algorithm with time complexity $O^*(2^{n/2}) \approx O^*(1.4145^n)$.

Drori & Peleg [9] use the technique of pruning the search tree to get a time complexity of $O^*(1.2494^n)$ for the Exact-Hitting-Set problem.

Exercise 52 *(Van Vliet [47])*
In the Three-Partition problem, the input consists of $3n$ positive integers $a_1, \ldots, a_n, b_1, \ldots, b_n,$ and $c_1, \ldots, c_n,$ together with an integer D. The question is to determine whether there exist three permutations π, ψ, ϕ of $\{1, \ldots, n\}$ such that $a_{\pi(i)} + b_{\psi(i)} + c_{\phi(i)} = D$ holds for all $i = 1, \ldots, n$. By checking all possible triples (π, ψ, ϕ) of permutations, this problem can be solved trivially in $O^*(n!^3)$ time.

Use the technique of preprocessing the data to improve the time complexity to $O^*(n!)$.

The binary knapsack problem. Here the input consists of n items that are specified by a positive integer value a_i and a positive integer weight w_i ($i = 1, \ldots, n$), together with a bound W. The goal is to find a subset of the items with the maximum total value subject to the condition that the total weight does not exceed W. The binary knapsack problem is closely related to the subset sum problem, and it can be solved (trivially) in $O^*(2^n)$ time. In 1974, Horowitz & Sahni [18] used a preprocessing trick to improve the time complexity to $O^*(2^{n/2})$.

For every $I \subseteq \{1, \ldots, \lfloor n/2 \rfloor\}$ we create a compound item x_I with value $a_I = \sum_{i \in I} a_i$ and weight $w_I = \sum_{i \in I} w_i$, and we put this item into the set X. For every $J \subseteq \{\lfloor n/2 \rfloor + 1, \ldots, n\}$ we put a corresponding compound item y_J into the set Y. The sets X and Y can be determined in $O^*(2^{n/2})$ time. The solution of the knapsack instance now reduces to the following: Find a compound item x_I in X and a compound item y_J in Y, such that $w_I + w_J \leq W$ and such that $a_I + a_J$ becomes maximum. But this can be handled by preprocessing as in our second toy problem, and we end up with an overall time complexity and an overall space complexity of $O^*(2^{n/2})$.

In 1981, Schroeppel & Shamir [45] improved the *space* complexity of this approach to $O^*(2^{n/4})$, while leaving its time complexity unchanged. The main trick is to split the instance into four pieces with $n/4$ items each, instead of two pieces with $n/2$ items. Apart from this, there has been no progress on exact algorithms for the knapsack problem since 1974.

Open problem 53 *Construct an exact algorithm for the subset sum problem or the knapsack problem with time complexity $O^*(c^n)$ for some $c < \sqrt{2}$, or prove that no such algorithm can exist under some reasonable complexity assumptions.*

6   Technique: Local Search

The idea of using local search methods in designing exact exponential time algorithms is relatively new. A *local search* algorithm is a search algorithm that wanders through the space of feasible solutions. At each step, this search algorithm moves from one feasible solution to another one nearby. In order to express the word 'nearby' mathematically, we need some notion of distance or neighborhood on the space of feasible solutions. For instance in the satisfiability problem, the feasible solutions are the truth assignments from the set X of logical variables to $\{0, 1\}$. A natural distance between truth assignments is the *Hamming* distance, that is, the number of bits where two truth assignments differ.

In this section we will concentrate on the 3-satisfiability problem where the input is a Boolean formula F in 3-CNF over the n logical variables in $X = \{x_1, x_2, \ldots, x_n\}$; see Section 4 for definitions and notations for this problem. We will describe three exact algorithms for 3-satisfiability that all are based on local search ideas. All three algorithms are centered around the Hamming neighborhood of truth assignments: For a truth assignment t and a non-negative integer d, we denote by $\mathcal{H}(t, d)$ the set of all truth assignments that have Hamming distance at most d from assignment t. It is easy to see that $\mathcal{H}(t, d)$ contains exactly $\sum_{k=0}^{d} \binom{n}{k}$ elements.

Exercise 61 *For a given truth assignment t and a given non-negative integer d, use the technique of pruning the search tree to check in $O^*(3^d)$ time whether the Hamming neighborhood $\mathcal{H}(t, d)$ contains a satisfying truth assignment for the 3-CNF formula F.*

In other words, the Hamming neighborhood $\mathcal{H}(t, d)$ can be searched quickly for the 3-satisfiability problem. For the k-satisfiability problem, the corresponding time complexity would be $O^*(k^d)$.

First local search approach to 3-satisfiability. We denote by 0^n (respectively, 1^n) the truth assignment that sets all variables to 0 (respectively, to 1). Any truth assignment is in $\mathcal{H}(0^n, n/2)$ or in $\mathcal{H}(1^n, n/2)$. Therefore by applying the search algorithm from Exercise 61 twice, we get an exact algorithm with running time $O^*(\sqrt{3}^n) \approx O^*(1.7321^n)$ for 3-satisfiability. It is debatable whether this algorithm should be classified under pruning the search tree or under local search. In any case, it is due to Schöning [44].

Second local search approach to 3-satisfiability. In the first approach, we essentially covered the whole solution space by two balls of radius $d = n/2$ centered at 0^n and 1^n. The second approach works with balls of radius $d = n/4$. The crucial idea is to *randomly* choose the center of a ball, and to search this ball with the algorithm from Exercise 61. If we only do this once, then we ignore most of the solution space, and the probability for answering correctly is pretty small. But by repeating this procedure a huge number α of times, we can boost the probability arbitrarily close to 1. A good choice for α is $\alpha = 100 \cdot 2^n / \sum_{k=0}^{n/4} \binom{n}{k}$. The algorithm now works as follows: Choose α times a truth assignment t uniformly at random, and search for a satisfying truth assignment in $\mathcal{H}(t, n/4)$. If in the end no satisfying truth assignment has been found, then answer that the formula F is not satisfiable.

We will now discuss the running time and the error probability of this algorithm. By Exercise 61, the running time can be bounded by roughly $\alpha \cdot 3^{n/4}$. By applying Stirling's approximation, one can show that up to a polynomial factor the expression $\sum_{k=0}^{n/4} \binom{n}{k}$ behaves asymptotically like $(256/27)^{n/4}$. Therefore, the upper bound $\alpha \cdot 3^{n/4}$ on the running time is in $O^*((3/2)^n) = O^*(1.5^n)$.

Now let us analyze the error probability of the algorithm. The only possible error occurs, if the formula F is satisfiable, whereas the algorithm does not manage to find a good ball $\mathcal{H}(t, n/4)$ that contains some satisfying truth assignment for F. For a single ball, the probability of containing a satisfying truth assignment equals $\sum_{k=0}^{n/4} \binom{n}{k}/2^n$, that is the number of elements in $\mathcal{H}(t, n/4)$ divided by the overall number of possible truth assignments. This probability equals $100/\alpha$. Therefore the probability of selecting a ball that does *not* contain any satisfying truth assignment is $1 - 100/\alpha$. The probability of α times *not* selecting such a ball equals $(1 - 100/\alpha)^\alpha$, which is bounded by the negligible value e^{-100}.

In fact, the whole algorithm can be derandomized without substantially increasing the running time. Dantsin, Goerdt, Hirsch, Kannan, Kleinberg, Papadimitriou, Raghavan & Schöning [6] do not choose the centers of the balls at random, but they take all centers from a so-called *covering code* so that the resulting balls cover the whole solution space. They show that such covering codes can be computed within reasonable amounts of time. The approach in [6] yields deterministic exact algorithms for k-satisfiability with running time $O^*((2 - \frac{2}{k+1})^n)$. For 3-satisfiability, [6] improve the time complexity further down to $O^*(1.4802^n)$ by using a smart idea for an underlying branching step. This is currently the fastest known deterministic algorithm for 3-satisfiability.

Third local search approach to 3-satisfiability. The first approach was based on selecting the center of a ball deterministically, and then searching through the whole ball. The second approach was based on selecting the center of a ball randomly, and then searching through the whole ball. The third approach now is based on selecting the center of a ball randomly, and then doing a short random walk within the ball. More precisely, the algorithm repeats the following procedure roughly $200 \cdot (4/3)^n$ times: Choose a truth assignment t uniformly at random, and perform $2n$ steps of a random walk starting in t. In each step, first select a violated clause at random, then select a literal in the selected clause at random, and finally flip the truth value of the corresponding variable. If in the very end no satisfying truth assignment has been found, then answer that the formula F is not satisfiable.

The intuition behind this algorithm is as follows. If we start far away from a satisfying truth assignment, then the random walk has little chance of stumbling towards a satisfying truth assignment. Hence, it is a good idea to terminate it quite early after $2n$ steps, without wasting time. But if the starting point is very close to a satisfying truth assignment, then the probability is high that the random walk will be dragged closer and closer towards this satisfying truth assignment. And if the random walk indeed is dragged into a satisfying truth assignment, then with high probability this happens within the first $2n$ steps of the random walk. The underlying mathematical structure is a Markov chain that can be analyzed by standard methods. Clearly, the error probability can be made negligibly small by sufficiently often restarting the random walk. And up to a polynomial factor, the running time of the algorithm is proportional to the number of performed random walks. This implies that the time complexity is $O^*((4/3)^n) \approx O^*(1.3334^n)$.

This algorithm and its analysis are due to Schöning [43]. Some of the underlying ideas go back to Papadimitriou [31] who showed that 2-SAT can be solved in polynomial time by a randomized local search procedure. The algorithm easily generalizes to the k-satisfiability problem, and yields a randomized exact algorithm with time complexity $O^*((2(k-1)/k)^n)$. The fastest known randomized exact algorithm for 3-satisfiability is due to Hofmeister, Schöning, Schuler & Watanabe [17], and has a running time of $O^*(1.3302^n)$. It is based on a refinement of the above random walk algorithm.

Open problem 62 *Design better deterministic and/or randomized algorithms for the k-satisfiability problem.*

More resuls on exact algorithms for k-satisfiability and related problems can be found in the work of Paturi, Pudlak & Zane [34], Paturi, Pudlak, Saks & Zane [35], Pudlak [37], Rodošek [40], and Williams [48].

7 How Can We Prove That a Problem Has No Sub-exponential Time Exact Algorithm?

All the problems discussed in this paper are NP-complete, and almost all of the developped algorithms use exponential time. Of course we cannot expect to find polynomial time algorithms for NP-complete problems, but maybe there exist better, sub-exponential, super-polynomial algorithms? How can we settle such questions?

Since our understanding of the landscape around the complexity classes P and NP still is fairly poor, the only available way of proving negative results on exact algorithms is by arguing relative to some widely believed conjectures. For instance, an NP-hardness proof establishes that some problem does not have a polynomial time algorithm, *given that the widely believed conjecture P$\neq$NP holds true*. The right conjecture for disproving the existence of sub-exponential time exact algorithms seems to be the following.

Widely believed conjecture 71 *SNP $\nsubseteq$ SUBEXP.*

We already mentioned in Section 2 that the class SNP is a broad complexity class that contains many important combinatorial optimization problems.

Therefore, if the widely believed Conjecture 71 is false, then quite unexpectedly all these important problems would possess relatively fast, sub-exponential time algorithms. However, the exact relationship between the P versus NP question and Conjecture 71 is unclear.

Open problem 72 *Does SNP ⊆ SUBEXP imply P = NP?*

Impagliazzo, Paturi & Zane [21] introduce the concept of *SERF-reduction* (Sub-Exponential Reduction Family) that preserves sub-exponential time complexities. Consider two problems A_1 and A_2 in NP with complexity parameters m_1 and m_2, respectively. A SERF-reduction from A_1 to A_2 is a family T_ε of Turing-reductions from A_1 to A_2 over all $\varepsilon > 0$ with the following two properties:

- The reduction $T_\varepsilon(x)$ can be done in time $\text{poly}(|x|) \cdot 2^{\varepsilon \cdot m_1(x)}$.
- If the reduction $T_\varepsilon(x)$ queries A_2 with input x', then $m_2(x')$ is linearly bounded in $m_1(x)$ and the length of x' is polynomially bounded in the length of x.

SERF-reducibility is transitive. Moreover, if problem A_1 is SERF-reducible to problem A_2 and if problem A_2 has a sub-exponential time algorithm, then also problem A_1 has a sub-exponential time algorithm. Consider some problem A that is hard for the complexity class SNP under SERF-reductions. If problem A had a sub-exponential time algorithm, then *all* the problems in SNP had sub-exponential time algorithms, and this would contradict the widely believed Conjecture 71 that SNP ⊄ SUBEXP.

The k-satisfiability problem plays a central role for sub-exponential time algorithms, the same central role that it plays everywhere else in computational complexity theory. There are two natural complexity parameters for k-satisfiability, the number of logical variables and the number of clauses. Impagliazzo, Paturi & Zane [21] prove that the two variants of k-satisfiability with these two complexity parameters are SERF-reducible to each other, and hence are equivalent under SERF-reductions. This indicates that for k-satisfiability the exact parameterization is not very important, and that all natural parameterizations of k-satisfiability should be SERF-reducible to each other. Most important, the paper [21] shows that for any fixed $k \geq 3$ the k-satisfiability problem is SNP-complete under SERF-reductions. As we discussed above, this implies that for any fixed $k \geq 3$ the k-satisfiability problem cannot have a sub-exponential time algorithm, unless SNP ⊆ SUBEXP. Therefore, the widely believed Conjecture 71 could also be formulated in the following way.

Widely believed conjecture 73 *(Exponential Time Hypothesis, ETH)*
For any fixed $k \geq 3$, k-satisfiability does not have a sub-exponential time algorithm.

Now let s_k denote the infimum of all real numbers δ with the property that there exists an $O^*(2^{\delta n})$ exact algorithm for solving the k-satisfiability problem. Observe that $s_k \leq s_{k+1}$ and $0 \leq s_k \leq 1$ hold trivially for all $k \geq 3$. The exponential time hypothesis conjectures $s_k > 0$ for all $k \geq 3$, and that the numbers s_k converge to some limit $s_\infty > 0$. Impagliazzo & Paturi [20] prove that under ETH, $s_k \leq (1 - \alpha/k) \cdot s_\infty$ holds, where α is some small positive constant. Consequently, under ETH we can never have $s_k = s_\infty$ and the time complexities for k-satisfiability must increase more and more as k increases.

Open problem 74 *(Impagliazzo & Paturi [20])*
Assuming the exponential time hypothesis for k-satisfiability, obtain evidence for the hypothesis that $s_\infty = 1$.

Now let us discuss the behavior of some other problems in NP. Impagliazzo, Paturi & Zane [21] show that for any fixed $k \geq 3$ the k-colorability problem is SNP-complete under SERF-reductions. Hence 3-colorability can not be solved in sub-exponential time, unless SNP $\subseteq$ SUBEXP. The paper [21] also shows that the Hamiltonian cycle problem and the independent set problem (both with the number of vertices as complexity parameter) can not be solved in sub-exponential time, unless SNP $\subseteq$ SUBEXP. Johnson & Szegedy [23] strengthen the result on the independent set problem by showing that the independent set problem in arbitrary graphs is equally difficult as in graphs with maximum degree three: Either both of these problems have a sub-exponential time algorithm, or neither of them does. Feige & Kilian [13] prove that also the bandwidth problem can not be solved in sub-exponential time, unless SNP $\subseteq$ SUBEXP. For all results listed in this paragraph, the proofs are done by translating classical NP-hardness proofs from the 1970s into SERF-reductions. The main technical problem is to keep the complexity parameters $m(x)$ under control.

In another line of research, Feige & Kilian [12] show that if in graphs with n vertices independent sets of size $O(\log n)$ can be found in polynomial time, then the 3-satisfiability problem can be solved in sub-exponential time. This result probably does not speak against the ETH, but indicates that finding small independent sets is difficult.

The W-hierarchy gives rise to yet another widely believed conjecture that can be used for disproving the existence of sub-exponential time exact algorithms. As we already mentioned in Section 2, the general belief is that all the W-classes are pairwise distinct. The following (cautious) conjecture only states that the W-hierarchy does not collapse completely.

Widely believed conjecture 75 *FPT $\neq$ W[P].*

Abrahamson, Downey & Fellows [1] proved that Conjecture 75 is false if and only if the satisfiability problem for Boolean circuits can be solved in sub-exponential time poly$(|x|) \cdot 2^{o(n)}$. Here $|x|$ denotes the size of the Boolean circuit that is given as an input, n denotes the number of input variables of the circuit, and $o(n)$ denotes some sub-linear function in n. Since the k-satisfiability problem is a special case of the Boolean circuit satisfiability problem, the exponential time hypothesis ETH implies Conjecture 75. It is not known whether the reverse implication also holds.

Most optimization problems that are mentioned in this survey possess exact algorithms with time complexity $O^*(c^{m(x)})$, i.e., exponential time where the exponent grows linearly in the complexity parameter $m(x)$. The quadratic assignment problem is a candidate for a natural problem that does not possess such an exact algorithm.

Open problem 76 *In the quadratic assignment problem (QAP) the input consists of two $n \times n$ matrices $A = (a_{ij})$ and $B = (b_{ij})$ ($1 \le i, j \le n$) with real entries. The objective is to find a permutation π that minimizes the cost function $\sum_{i=1}^{n} \sum_{j=1}^{n} a_{\pi(i)\pi(j)} b_{ij}$. The QAP can be solved in $O^*(n!)$ time. The QAP is a notoriously hard problem, and no essentially faster algorithms are known (Pardalos, Rendl & Wolkowicz [33]).*

Prove that (under some reasonable complexity assumptions) the QAP can not be solved in $O^*(c^n)$ time, for any fixed value c.

8   Concluding Remarks

Currently, when we are dealing with an optimization problem, we are used to look at its computational complexity, its approximability behavior, its online behavior (with respect to competitive analysis), its polyhedral structure. Exact algorithms with good worst case behavior should probably become another standard item on this list, and we feel that the known techniques and results as described in Sections 3–6 deserve to be taught in our introductory algorithms courses.

There remain many open problems and challenging questions around the worst case analysis of exact algorithms for NP-hard problems. This seems to be a rich and promising area. We only have a handful of techniques available, and there is ample space for improvements and for new results.

Acknowledgement. I thank David Eppstein, Jesper Makholm Nielsen, and Ryan Williams for several helpful comments on preliminary versions of this paper, and for providing some pointers to the literature. Furthermore, I thank an unknown referee for many suggestions how to improve the structure, the English, the style, and the contents of this paper.

References

1. K.A. ABRAHAMSON, R.G. DOWNEY, AND M.R. FELLOWS [1995]. Fixed-parameter tractability and completeness IV: On completeness for W[P] and PSPACE analogues. *Annals of Pure and Applied Logic 73*, 235–276.

2. D. APPLEGATE, R. BIXBY, V. CHVÁTAL, AND W. COOK [1998]. On the solution of travelling salesman problems. *Documenta Mathematica 3*, 645–656.

3. R. BEIGEL [1999]. Finding maximum independent sets in sparse and general graphs. *Proceedings of the 10th ACM-SIAM Symposium on Discrete Algorithms (SODA'1999)*, 856–857.

4. R. BEIGEL AND D. EPPSTEIN [1995]. 3-Coloring in time $O(1.3446^n)$: A no-MIS algorithm. *Proceedings of the 36th Annual Symposium on Foundations of Computer Science (FOCS'1995)*, 444–453.

5. J. CHEN, I.A. KANJ, AND W. JIA [1999]. Vertex cover: Further observations and further improvements. *Proceedings of the 25th Workshop on Graph Theoretic Concepts in Computer Science (WG'1999)*, Springer, LNCS 1665, 313–324.

6. E. DANTSIN, A. GOERDT, E.A. HIRSCH, R. KANNAN, J. KLEINBERG, C.H. PAPADIMITRIOU, P. RAGHAVAN, AND U. SCHÖNING [2001]. A deterministic $(2 - \frac{2}{k+1})^n$ algorithm for k-SAT based on local search. To appear in *Theoretical Computer Science*.

7. R.G. DOWNEY AND M.R. FELLOWS [1992]. Fixed parameter intractability. *Proceedings of the 7th Annual IEEE Conference on Structure in Complexity Theory (SCT'1992)*, 36–49.

8. R.G. DOWNEY AND M.R. FELLOWS [1999]. *Parameterized complexity*. Springer Monographs in Computer Science.

9. L. DRORI AND D. PELEG [1999]. Faster exact solutions for some NP-hard problems. *Proceedings of the 7th European Symposium on Algorithms (ESA'1999)*, Springer, LNCS 1643, 450–461.

10. D. EPPSTEIN [2001]. Improved algorithms for 3-coloring, 3-edge-coloring, and constraint satisfaction. *Proceedings of the 12th ACM-SIAM Symposium on Discrete Algorithms (SODA'2001)*, 329–337.

11. D. EPPSTEIN [2001]. Small maximal independent sets and faster exact graph coloring. *Proceedings of the 7th Workshop on Algorithms and Data Structures (WADS'2001)*, Springer, LNCS 2125, 462–470.

12. U. FEIGE AND J. KILIAN [1997]. On limited versus polynomial nondeterminism. *Chicago Journal of Theoretical Computer Science* (http://cjtcs.cs.uchicago.edu/).

13. U. FEIGE AND J. KILIAN [2000]. Exponential time algorithms for computing the bandwidth of a graph. Manuscript.

14. M.R. GAREY AND D.S. JOHNSON [1979]. *Computers and Intractability: A Guide to the Theory of NP-Completeness*. Freeman, San Francisco.

15. J. GRAMM AND R. NIEDERMEIER [2000]. Faster exact solutions for Max2Sat. *Proceedings of the 4th Italian Conference on Algorithms and Complexity (CIAC'2000)*, Springer, LNCS 1767, 174–186.

16. M. HELD AND R.M. KARP [1962]. A dynamic programming approach to sequencing problems. *Journal of SIAM 10*, 196–210.

17. T. HOFMEISTER, U. SCHÖNING, R. SCHULER, AND O. WATANABE [2001]. A probabilistic 3-SAT algorithm further improved. Manuscript.

18. E. HOROWITZ AND S. SAHNI [1974]. Computing partitions with applications to the knapsack problem. *Journal of the ACM 21*, 277–292.

19. R.Z. HWANG, R.C. CHANG, AND R.C.T. LEE [1993]. The searching over separators strategy to solve some NP-hard problems in subexponential time. *Algorithmica 9*, 398–423.

20. R. IMPAGLIAZZO AND R. PATURI [2001]. Complexity of k-SAT. *Journal of Computer and System Sciences 62*, 367–375.

21. R. IMPAGLIAZZO, R. PATURI, AND F. ZANE [1998]. Which problems have strongly exponential complexity? *Proceedings of the 39th Annual Symposium on Foundations of Computer Science (FOCS'1998)*, 653–663.

22. T. JIAN [1986]. An $O(2^{0.304n})$ algorithm for solving maximum independent set problem. *IEEE Transactions on Computers 35*, 847–851.

23. D.S. JOHNSON AND M. SZEGEDY [1999]. What are the least tractable instances of max independent set? *Proceedings of the 10th ACM-SIAM Symposium on Discrete Algorithms (SODA'1999)*, 927–928.

24. O. KULLMANN [1997]. Worst-case analysis, 3-SAT decisions, and lower bounds: Approaches for improved SAT algorithms. In: *The Satisfiability Problem: Theory and Applications*, D. Du, J. Gu, P.M. Pardalos (eds.), DIMACS Series in Discrete Mathematics and Theoretical Computer Science 35, 261–313.

25. O. KULLMANN [1999]. New methods for 3-SAT decision and worst case analysis. *Theoretical Computer Science 223*, 1–72.

26. E.L. LAWLER [1976]. A note on the complexity of the chromatic number problem. *Information Processing Letters 5*, 66–67.

27. R.J. LIPTON AND R.E. TARJAN [1979]. A separator theorem for planar graphs. *SIAM Journal on Applied Mathematics 36*, 177–189.

28. B. MONIEN AND E. SPECKENMEYER [1985]. Solving satisfiability in less than 2^n steps. *Discrete Applied Mathematics 10*, 287–295.

29. J.W. MOON AND L. MOSER [1965]. On cliques in graphs. *Israel Journal of Mathematics 3*, 23–28.

30. J.M. NIELSEN [2001]. Personal communication.

31. C.H. PAPADIMITRIOU [1991]. On selecting a satisfying truth assignment. *Proceedings of the 32nd Annual Symposium on Foundations of Computer Science (FOCS'1991)*, 163–169.

32. C.H. PAPADIMITRIOU AND M. YANNAKAKIS [1991]. Optimization, approximation, and complexity classes. *Journal of Computer and System Sciences 43*, 425–440.

33. P. PARDALOS, F. RENDL, AND H. WOLKOWICZ [1994]. The quadratic assignment problem: A survey and recent developments. In: *Proceedings of the DIMACS Workshop on Quadratic Assignment Problems*, P. Pardalos and H. Wolkowicz (eds.), DIMACS Series in Discrete Mathematics and Theoretical Computer Science 16, 1–42.

34. R. PATURI, P. PUDLAK, AND F. ZANE [1997]. Satisfiability coding lemma. *Proceedings of the 38th Annual Symposium on Foundations of Computer Science (FOCS'1997)*, 566–574.

35. R. PATURI, P. PUDLAK, M.E. SAKS, AND F. ZANE [1998]. An improved exponential time algorithm for k-SAT. *Proceedings of the 39th Annual Symposium on Foundations of Computer Science (FOCS'1998)*, 628–637.

36. M. PAULL AND S. UNGER [1959]. Minimizing the number of states in incompletely specified sequential switching functions. *IRE Transactions on Electronic Computers 8*, 356–367.

37. P. PUDLAK [1998]. Satisfiability – algorithms and logic. *Proceedings of the 23rd International Symposium on Mathematical Foundations of Computer Science (MFCS'1998)*, Springer, LNCS 1450, 129–141.

38. J.M. ROBSON [1986]. Algorithms for maximum independent sets. *Journal of Algorithms 7*, 425–440.

39. J.M. ROBSON [2001]. Finding a maximum independent set in time $O(2^{n/4})$? Manuscript.

40. R. RODOŠEK [1996]. A new approach on solving 3-satisfiability. *Proceedings of the 3rd International Conference on Artificial Intelligence and Symbolic Mathematical Computation* Springer, LNCS 1138, 197–212.

41. I. SCHIERMEYER [1992]. Solving 3-satisfiability in less than $O(1.579^n)$ steps. *Selected papers from Computer Science Logic (CSL'1992)*, Springer, LNCS 702, 379–394.

42. I. SCHIERMEYER [1993]. Deciding 3-colorability in less than $O(1.415^n)$ steps. *Proceedings of the 19th Workshop on Graph Theoretic Concepts in Computer Science (WG'1993)*, Springer, LNCS 790, 177–182.

43. U. SCHÖNING [1999]. A probabilistic algorithm for k-SAT and constraint satisfaction problems. *Proceedings of the 40th Annual Symposium on Foundations of Computer Science (FOCS'1999)*, 410–414.
44. U. SCHÖNING [2001]. New algorithms for k-SAT based on the local search principle. *Proceedings of the 26th International Symposium on Mathematical Foundations of Computer Science (MFCS'2001)*, Springer, LNCS 2136, 87–95.
45. R. SCHROEPPEL AND A. SHAMIR [1981]. A $T = O(2^{n/2})$, $S = O(2^{n/4})$ algorithm for certain NP-complete problems. *SIAM Journal on Computing 10*, 456–464.
46. R.E. TARJAN AND A.E. TROJANOWSKI [1977]. Finding a maximum independent set. *SIAM Journal on Computing 6*, 537–546.
47. A. VAN VLIET [1995]. Personal communication.
48. R. WILLIAMS [2002]. Algorithms for quantified Boolean formulas. *Proceedings of the 13th ACM-SIAM Symposium on Discrete Algorithms (SODA'2002)*.

Author Index

Lecture Notes in Computer Science

For information about Vols. 1–2494

please contact your bookseller or Springer-Verlag